METHODS OF SOLVING SINGULAR SYSTEMS OF ORDINARY DIFFERENTIAL EQUATIONS

Yuri Boyarintsev
Irkutsk Computer Centre, USSR Academy of Sciences

Translated by

Vassily Michalkowski
Irkutsk Computer Centre, USSR Academy of Sciences

A Wiley–Interscience Publication

JOHN WILEY & SONS
Chichester · New York · Brisbane · Toronto · Singapore

Originally published in 1988 under the title *Metody resheniya vyrozhdennykh sistem obyknovennykh differentsial'nykh uravneniy* by Nauka Publishing House, Siberian Division

Other Wiley Editorial Offices

John Wiley & Sons, Inc., 605 Third Avenue,
New York, NY 10158–0012, USA

Jacaranda Wiley Ltd, G.P.O. Box 859, Brisbane,
Queensland 4001, Australia

John Wiley & Sons (Canada) Ltd, 5353 Dundas Street West, Fourth Floor,
Etobicoke, Ontario M9B-6H8, Canada

John Wiley & Sons (SEA) Pte Ltd, 37 Jalan Pemimpin 05-04,
Block B, Union Industrial Building, Singapore 2057

Library of Congress Cataloging-in-Publication Data:
Boiarintsev, IU. E. (IUriĭ Eremeevich)
    [Metody resheniia vyrozhennykh sistem obyknovennykh
differentsial 'nykh uravneiĭ. English]
    Methods of solving singular systems of ordinary differential
equations / Yuri Boyarintsev ; translated by Vassily Michalkowski.
        p.   cm. — (Pure and applied mathematics)
    Translation of: Metody resheniia vyrozhennykh sistem obyknovennykh
differentsial 'nykh uravneiĭ.
    "A Wiley–Interscience Publication" — T.p.
    Includes bibliographical references and index.
    ISBN 0 471 93163 2
    1. Differential equations—Numerical solutions. 2. Equations,
Simultaneous—Numerical solutions. I. Title. II. Series: Pure and
applied mathematics (John Wiley & Sons : Unnumbered)
QA372.B664   1992
515'.35—dc20   91-36823
                                                             CIP

*A catalogue record for this book is
available from the British Library.*

ISBN 0 471 93163 2

Typeset by Thomson Press (India) Ltd, New Delhi, India
Printed in Great Britain by Biddles Ltd, Guildford

CONTENTS

FOREWORD

Twelve years have now elapsed since the appearance of my monograph *Regular and Singular Systems of Linear Ordinary Differential Equations* (Nauka, Novosibirsk, 1980). During that period it has been possible to comprehend the results presented in that book as regards applications and to learn to apply them to construct stable difference and other approximations to singular systems of ordinary differential equations. As a result, material emerged for a new monograph which is here offered for the reader's attention.

In Chapter 1, on the basis of different generalized inverse matrices, a study is made of pairs and triplets of matrices accompanying singular systems. The results presented in this chapter, taken together, represent an algebraic apparatus to be used throughout the book. In the first reading, however, a detailed study of Chapter 1 is not obligatory: it is sufficient to familiarize oneself only with the formulations of the theorems, lemmas, and corollaries.

Chapter 2 identifies classes of singular systems, the solutions of which are reduced to solving the simplest index-1 systems, and converging difference schemes are constructed and substantiated for each of the classes of systems identified. Note that the validity of the schemes obtained was confirmed by computer simulations.

Chapter 3 is devoted to the derivation of formulae for general solutions of the systems identified in Chapter 2 (with additional conditions admitting the notation in the form of Stieltjes integrals). These formulae are used to prove the theorems of existence and uniqueness of the solutions. In addition, it is demonstrated that, when constructing difference schemes, the formulae obtained rapidly meet with success.

The book gives an account of the present status of the problem and a review of the latest publications.

A number of valuable remarks, which helped to improve the manuscript, were made by Corresponding Member of the USSR Academy of Sciences, V. M. Matrosov, and by Drs V. P. Bulatov, R. I. Kozlov, V. F. Chistyakov, and V. A. Danilov. The author expresses his profound gratitude to all of them.

1 GENERALIZED INVERSE MATRICES AND SOME OF THEIR APPLICATIONS

This chapter presents results on so-called generalized inverse matrices (semi-inverse and pseudo-inverse matrices and inverse Drasin matrices) which, taken together, represent an algebraic apparatus to be used in subsequent chapters, when studying singular systems of ordinary differential equations and their difference approximations.

1.1. Semi-inverse matrices

Let us start with a definition.

Definition 1.1.1 A semi-inverse matrix for an $(m \times n)$ matrix A is said to be an $(n \times m)$ matrix X that satisfies the equation

$$AXA = A. \tag{1.1.1}$$

A semi-inverse matrix exists for any $(m \times n)$ matrix (see [1, p. 32]). However, it is generally not unique.

In what follows, each of the semi-inverse matrices for the matrix A will be denoted by A^- so that

$$AA^-A = A. \tag{1.1.2}$$

A description of the set of semi-inverse matrices for the matrix A is given by the following lemmas.

Lemma 1.1.1 A general solution of the matrix equation (1.1.1) is representable as

$$X = A^- + (E - A^- A)U + V(E - AA^-), \qquad (1.1.3)$$

where A^- is any semi-inverse matrix for the matrix A, and U and V are arbitrary matrices of appropriate dimensions.

In other words, by knowing one semi-inverse matrix for the matrix A, all the other matrices are obtainable from formula (1.1.3).

Proof It is obvious that a matrix having the form (1.1.3) satisfies equation (1.1.1). Conversely, if the matrix X satisfies equation (1.1.1), then it is representable as (1.1.3) when $U = X$ and $V = A^- AX - A^-$. This completes the proof of Lemma 1.1.1.

Lemma 1.1.2 Let $A_1 = PAQ$, where the matrices P and Q are nonsingular. Then if the matrix X runs through all semi-inverse matrices for the matrix A, then the matrix $X_1 = Q^{-1}XP^{-1}$ runs through all semi-inverse matrices for the matrix A_1.*

Proof Let A^- be a semi-inverse matrix for the matrix A. Then the matrix $A_1^- = Q^{-1}A^-P^{-1}$ is a semi-inverse matrix for the matrix A_1. Indeed,

$$A_1 A_1^- A_1 = PAQQ^{-1}A^-P^{-1}PAQ = PAA^- AQ = PAQ = A_1.$$

If, however, the matrix A_1^- is semi-inverse for the matrix A_1, then it is, of necessity, representable as $A_1^- = Q^{-1}A^-P^{-1}$, where the matrix $A^- = QA_1^-P$, as a consequence of the equalities

$$AA^- A = AQA_1^- PA = P^{-1}PAQA_1^- PAQQ^{-1} = P^{-1}A_1 A_1^- A_1 Q^{-1}$$
$$= P^{-1}A_1 Q^{-1} = P^{-1}PAQQ^{-1} = A,$$

is semi-inverse for the matrix A. The lemma is proved.

The following lemmas facilitate the determination of the semi-inverse matrices.

Lemma 1.1.3 The block matrix (R, S), in which

$$S = (E - A^- A)(B - BA^- A)^-, \qquad R = A^- - SBA^-,$$

is a semi-inverse matrix for the block matrix $\begin{pmatrix} A \\ B \end{pmatrix}$.

---

*In other words, the matrix $X_1 = Q^{-1}XP^{-1}$ is semi-inverse for $A_1 = PAQ$ if and only if X is semi-inverse for A.

Lemma 1.1.4 The block matrix $\begin{pmatrix} R \\ S \end{pmatrix}$, in which

$$S = (B - AA^- B)^- (E - AA^-),$$
$$R = A^- - A^- BS,$$

is a semi-inverse matrix for the block matrix (A, B).

It is easy to prove Lemmas 1.1.3 and 1.1.4 by simply verifying the validity of the equalities

$$\begin{pmatrix} A \\ B \end{pmatrix} (R, S) \begin{pmatrix} A \\ B \end{pmatrix} = \begin{pmatrix} A \\ B \end{pmatrix}, \qquad (A, B) \begin{pmatrix} R \\ S \end{pmatrix} (A, B) = (A, B),$$

and the assertions of the formulated lemmas make it possible to construct numerous algorithms to calculate semi-inverse matrices. Let, for example, the $(i \times n)$ matrix A_i be formed by the first i rows of the $(m \times n)$ matrix A. Row k $(1 \leqslant k \leqslant m)$ in the matrix A will be denoted by $a_{k.}$. Then on the basis of Lemma 1.1.3 the following recurrent process is obtained:

$$A_{i+1}^- = (R_{i+1}, S_{i+1}),$$

where

$$S_{i+1} = (E - A_i^- A_i)(a_{i+1} - a_{i+1} A_i^- A_i)^-,$$
$$R_{i+1} = A_i^- - S_{i+1} a_{i+1} A_i^-, \qquad i = 1, \ldots, m-1,$$

and since $A_m = A$, the last step of this recurrent process will result in a semi-inverse matrix for the matrix A. In this case it turns out that at each step one has to find semi-inverse matrices only for one-row matrices.

It is particularly easy to specify one of the semi-inverse matrices for a one-row matrix

$$A = (\alpha_1, \alpha_2, \ldots, \alpha_n) \tag{1.1.4}$$

if all α_i are zero, then

$$A^- = \begin{pmatrix} 0 \\ 0 \\ \vdots \\ 0 \end{pmatrix}. \tag{1.1.5}$$

Otherwise, one takes any nonzero element α_i (perhaps, being a maximum in absolute value) and it is assumed that

$$A^- = \begin{pmatrix} 0 \\ \vdots \\ \alpha_i^{-1} \\ \vdots \\ 0 \end{pmatrix} \tag{1.1.6}$$

(the matrix (1.1.6) involves only one nonzero element which is located in row i and equals α_i^{-1}).

It is easy to check that the matrix A^- constructed in this way (see (1.1.5) and (1.1.6)) is, indeed, semi-inverse for the one-row matrix (1.1.4). Hence, if the possible instability to round-off errors is disregarded, one obtains a rather simple algorithm for calculating semi-inverse matrices.

From Lemmas 1.1.3 and 1.1.4 some useful corollaries follow.

Corollary 1.1.1 In the block matrix

$$\begin{pmatrix} A \\ B \end{pmatrix} \tag{1.1.7}$$

let the submatrix A be nonsingular. Then the block matrix of the form

$$(A^{-1}, \quad 0)$$

is a semi-inverse matrix for the matrix (1.1.7) (here 0 is a zero matrix).

Proof Since the matrix A^{-1} is the only semi-inverse matrix for the nonsingular matrix A, then in the formulation of Lemma 1.1.3, $S = 0$ and $R = A^{-1}$ and the assertion being proved is an obvious one.

Corollary 1.1.2 In the block matrix

$$(A, B) \tag{1.1.8}$$

let the submatrix A be nonsingular. Then the block matrix

$$\begin{pmatrix} A^{-1} \\ 0 \end{pmatrix}$$

is a semi-inverse matrix for the matrix (1.1.8).

The proof is analogous to the previous one.

Corollary 1.1.3 Let a block representation of the $(m \times n)$ matrix A have the form

$$A = \begin{pmatrix} C \\ D \end{pmatrix}, \tag{1.1.9}$$

where $C = (A_{11}, A_{12})$ and $D = (A_{21}, A_{22})$, with the matrix A_{11} being nonsingular and its order $r < \min(m, n)$ coinciding with the rank of the matrix A, i.e. rank A = rank A_{11}. Then the matrix

$$A^- = \begin{pmatrix} A_{11}^{-1} & 0 \\ 0 & 0 \end{pmatrix}$$

is one of the semi-inverse matrices for the matrix A.

Proof For the proof, we shall use the following well-known result.

Lemma 1.1.5 [1] If in the block representation of the matrix A

$$A = \begin{pmatrix} A_{11} & A_{12} \\ A_{21} & A_{22} \end{pmatrix}$$

the matrix A_{11} is nonsingular, then in order that the equality of ranks rank $A = $ rank A_{11} should hold, it is necessary and sufficient that the block A_{22} has the representation

$$A_{22} = A_{21}A_{11}^{-1}A_{12}.$$

This result, with the conditions formulated, allows us to conclude that the matrix D in (1.1.9) has the form

$$D = (A_{21}, A_{21}A_{11}^{-1}A_{12}).$$

But then, by applying Lemma 1.1.3 and Corollary 1.1.1 we arrive at the proof of Corollary 1.1.3. Indeed, by virtue of Corollary 1.1.1

$$\begin{pmatrix} A_{11}^{-1} \\ 0 \end{pmatrix} = C^-,$$

and therefore

$$E - C^-C = \begin{pmatrix} 0 & -A_{11}^{-1}A_{12} \\ 0 & E \end{pmatrix}, \qquad D - DC^-C = (0,0), \qquad (1.1.10)$$

and since a zero matrix can be taken as the semi-inverse matrix for the matrix (1.1.0), then for the matrices S and R from Lemma 1.1.3 we obtain the values

$$S = 0, \qquad R = C^-.$$

But then, according to Lemma 1.1.3, we have

$$(R, S) = \begin{pmatrix} A_{11}^{-1} & 0 \\ 0 & 0 \end{pmatrix} = A^- \qquad (1.1.11)$$

and Corollary 1.1.3 is proved.

Using Corollaries 1.1.1–1.1.3 one can construct a semi-inverse matrix for any matrix A. To accomplish this, it is sufficient to rearrange the rows and columns of the matrix A to obtain the matrix A_1 allowing for the application of one of Corollaries 1.1.1–1.1.3 to it. From algebra it is known that such rearrangements are feasible and that the equality

$$A = PA_1Q$$

holds, where P and Q are matrices of rearrangements of corresponding rows and columns, and $P^*P = PP^* = E$ and $Q^*Q = QQ^* = E$, i.e. $P^{-i} = P^*$ and $Q^{-i} = Q^*$.

But then, having obtained, with the help of Corollaries 1.1.1–1.1.3 the semi-inverse matrix A_1^- for the matrix A_1, on the basis of Lemma 1.1.2 we have

$$A^- = Q^* A_1 P^*. \tag{1.1.12}$$

It is interesting to note that the matrices A_1^- and A^- in (1.1.12) satisfy two equations:

$$AXA = A, \qquad XAX = X. \tag{1.1.13}$$

In what follows, each of the semi-reciprocal matrices for the matrix A satisfying the system (1.1.13) will be referred to as the inverse semi-reciprocal matrix for the matrix A and will be denoted by $A^\sim$ such that

$$AA^\sim A = A, \qquad A^\sim AA^\sim = A^\sim. \tag{1.1.14}$$

A description of the set of inverse semi-reciprocal matrices is given by the following lemmas.

Lemma 1.1.6   A general solution of the system of matrix equations (1.1.13) can be represented as

$$X = A^- AA^- + A^- AU(E - AA^-) + (E - A^- A)UAA^-$$
$$+ (E - A^- A)UAU(E - AA^-), \tag{1.1.15}$$

where A^- is any semi-reciprocal matrix for the matrix A, and U is an arbitrary matrix of suitable dimensions. In other words, using the formula (1.1.15) one may obtain all inverse semi-reciprocal matrices for the matrix A.

Proof   By means of direct substitution it is easy to check that the matrix having the form (1.1.15) satisfies the system (1.1.13). Conversely, if the matrix X satisfies the system (1.1.13), then it is representable as (1.1.15) when $U = X - A^- AA^-$. The proof of Lemma 1.1.6 is thereby completed.

Lemma 1.1.7   When the matrix X has run through all semi-inverse matrices for the matrix A, the matrix $Y = XAX$ will run through all inverse semi-reciprocal matrices for the matrix A.

Proof   Let X be a semi-inverse matrix for the matrix A, i.e. $AXA = A$. Then

$$AYA = AXAXA = AXA = A,$$
$$YAY \neq X(AXA)XAX = X(AXA)X = XAX = Y.$$

Thus, the matrix $Y = XAX$ satisfies the system (1.1.13) and is, therefore, an inverse semi-reciprocal matrix for the matrix A. If, however, Y is an arbitrary inverse semi-reciprocal matrix for the matrix A, then by virtue of the second equation of the system (1.1.13) it is representable as $Y = XAX$, with the value of the semi-inverse matrix X for the matrix A equal to the matrix Y itself. The lemma is proved.

Finally, let us prove one more useful lemma.

Lemma 1.1.8 If X runs through all semi-inverse matices for the matrix A, then X^* runs through all semi-inverse matrices for the matrix A^*.

Proof If $AXA = A$, then $A^*X^*A^* = A^*$; therefore, if X is a semi-inverse matrix for the matrix A, then X^* is a semi-inverse matrix for the matrix A^*. Just as above, if Y is a semi-inverse matrix for the matrix A^*, then Y^* is a semi-inverse matrix for the matrix A. Hence, for any semi-inverse matrix Y for the matrix A^*, there will be a semi-inverse matrix for the matrix A, a conjugate one which coincides with the matrix Y. Note further that the mapping $X \rightarrow X^*$ is bijective because $X_1^* \neq X_2^*$ follows from $X_1 \neq X_2$.

Bearing in mind this lemma as well as the fact that the symbol $^-$ denotes any semi-inverse matrix for the matrix $\mathscr{M}$, we shall subsequently write

$$(A^-)^* = (A^*)^-, \qquad (A^*)^- = (A^-)^*. \tag{1.1.16}$$

1.2. The pseudo-inverse matrix

Among inverse semi-reciprocal matrices, the so-called pseudo-inverse matrix has a special place. The exclusiveness of the pseudo-inverse matrix is attributed to the fact that it exists for every matrix, is unique, and permits the problem of least squares to be solved for linear algebraic systems (see [1, p. 32] as well as [2, p. 40]).

Before introducing the definition, it must be recalled that an arbitrary rectangular $(m \times n)$ matrix A having rank r is representable as the product $A = B \cdot C$ of the matrices B and C, respectively, of dimensions $(m \times r)$ and $(r \times n)$. This representation is called the skeleton expansion of the matrix A (see [1, p. 32]). It is not unique; however, the matrices B^*B and CC^* are nonsingular and the equality of ranks $\operatorname{rank} B = \operatorname{rank} C = \operatorname{rank} A$ holds.

Definition 1.2.1 The matrix A^+ of dimensions $(n \times m)$ is called the pseudo-inverse matrix to the $(m \times n)$ matrix A if the equalities

(1) $AA^+A = A$,
(2) $A^+AA^+ = A^+$,
(3) $(AA^+)^* = AA^+$,
(4) $(A^+A)^* = A^+A$

$$\tag{1.2.1}$$

are satisfied (see, for example, [3]).

The following theorem holds true.

Theorem 1.2.1 A pseudo-inverse matrix to an arbitrary matrix A exists, is unique, and has the representation

$$A^+ = C^*(CC^*)^{-1}(B^*B)^{-1}B^*, \qquad (1.2.2)$$

where B and C are the components of the skeleton expansion $A = B \cdot C$ matrix A.

Proof The fact that a pseudo-inverse matrix for the matrix exists and is representable as (1.2.2) is proved in a straightforward way through direct verification of the fulfilment of the equalities (1.2.1) for the matrix (1.2.2). It remains to prove that two different pseudo-inverse matrices A_1^+ and A_2^+ cannot exist for a given matrix A.

Indeed, let

$$\mathscr{D} = A_1^+ - A_2^+.$$

Then the first equality of (1.2.1) yields

$$A\mathscr{D}A = 0, \qquad (1.2.3)$$

and the third and fourth equalities give

$$(\mathscr{D}A)^* = \mathscr{D}A, \qquad (A\mathscr{D})^* = A\mathscr{D}. \qquad (1.2.4)$$

The equalities (1.2.3) and (1.2.4) give the equalities

$$(\mathscr{D}A)^*\mathscr{D}A = (\mathscr{D}A)^2 = \mathscr{D}A\mathscr{D}A = 0$$
$$(A\mathscr{D})^*A\mathscr{D} = (A\mathscr{D})^2 = A\mathscr{D}A\mathscr{D} = 0$$

and, consequently, $A\mathscr{D} = 0$ and $\mathscr{D}A = 0$ because $\mathscr{M} = 0$ follows from $\mathscr{M}^*\mathscr{M} = 0$ for any matrix $\mathscr{M}$. But then

$$A_1^+A = A_2^+A, \qquad AA_1^+ = AA_2^+$$

and, in view of (2) from (1.2.1), we have

$$A_1^+ = A_1^+AA_1^+ = A_2^+AA_1^+ = A_2^+AA_2^+ = A_2^+.$$

Theorem 1.2.1 is proved.

Corollary 1.2.1 If the columns of the $(m \times n)$ matrix A are linearly independent, i.e. rank $A = n$, then

$$A^+ = (A^*A)^{-1}A^*. \qquad (1.2.5)$$

Proof Under the condition of Corollary 1.2.1 the skeleton expansion $A = BC$ of the matrix A is thus

$$A = A \cdot E.$$

But then (1.2.5) follows from (1.2.2).

Corollary 1.2.2 If the rows of the $(m \times n)$ matrix A are linearly independent, i.e. rank $A = m$, than

$$A^+ = A^*(AA^*)^{-1}. \tag{1.2.6}$$

The proof is analogous to the previous one.

If the elements of the matrix A are functions of the parameter $t \in [\alpha, \beta]$, i.e. $A = A(t)$, then one can formulate the problem of finding the derivative of the matrix $A^+ = A^+(t)$. In this case the following theorem is valid.

Theorem 1.2.2 If $A \in C^1[\alpha, \beta]$, $A^+ \in C^1[\alpha, \beta]$, then

$$(A^+)' = -A^+A'A^+ + (E - A^+A)(A^+A')^*A^+ \\ + A^*(A'A^+)^*(E - AA^+).$$

Proof In view of the equality (2) from (1.2.1), we obtain:

$$(A^+)' = (A^+AA^+)' = (A^+A)'A^+ + A^+A(A^+)' + A^+A'A^+ \\ - A^+A'A^+ = (A^+A)'A^+ + A^+(AA^+)' - A^+A'A^+,$$

i.e.

$$(A^+)' = -A^+A'A^+ + (A^+A)'A^+ + A^+(AA^+)'. \tag{1.2.7}$$

Besides, since

$$A(E - A^+A) = 0, \qquad (A^+A)^* = A^+A,$$

we have

$$(E - A^+A)A^* = [[(E - A^+A)A^*]^*]^* = (A(E - A^+A))^* = 0,$$

i.e. $(E - A^+A)A^* = 0$; therefore

$$0 = [(E - A^+A)A^*]' = -(A^+A)'A^* + (E - A^+A)(A')^*$$

and, consequently,

$$(A^+A)'A^* = (E - A^+A)(A')^*. \tag{1.2.8}$$

Now, we note that, since $(A^+A)^* = A^+A$ and $A^+AA^+ = A^+$, the following equality holds:

$$A^*(A^+)^*A^+ = [[A^*(A^+)^*A^+]^*]^* = (A^+)^*A^+A^* = A^+AA^+ = A^+. \tag{1.2.9}$$

Moreover,

$$(A')^*(A^+)^*A^+ = [[(A')^*(A^+)^*A^+]^*]^* = [(A^+)^*A^+A']^* = (A^+A')^*A^+. \tag{1.2.10}$$

But then, by multiplying the equality (1.2.8) by the matrix $(A^+)^*A^+$ on the right, in view of (1.2.9) and (1.2.10), we obtain:

$$(A^+A)'A^+ = (E - A^+A)(A^+A')^*A^+. \tag{1.2.11}$$

In much the same way, the equality

$$A^+(AA^+)' = A^+(A'A^+)^*(E - AA^+) \tag{1.2.12}$$

is derived, and substitution of (1.2.11) and (1.2.12) into the equality (1.2.7) leads to formula (1.2.6). Theorem 1.2.2 is proved.

Corollary 1.2.3 If $A \in C^1[\alpha, \beta]$ and $A^+ \in C^1[\alpha, \beta]$, then

$$(A^+A)' = A^+A'(E - A^+A) + (E - A^+A)(A^+A')^*, \tag{1.2.13}$$

$$(AA^+)' = (E - AA^+)A'A^+ + (A'A^+)^*(E - AA^+). \tag{1.2.14}$$

Proof In view of the equality (1.2.6), we obtain:

$$\begin{aligned}(A^+A)' &= (A^+)'A + A^+A' \\ &= A^+A'(E - A^+A) + (E - A^+A)(A^+A')^*A^+A, \end{aligned} \tag{1.2.15}$$

which coincides with formula (1.2.13) because in (1.2.15)

$$(A^+A')^*A^+A = [[(A^+A')^*A^+A]^*]^* = [A^+AA^+A']^* = (A^+A')^*.$$

The equality (1.2.14) is proved along similar lines.

Finally, we must note the following two properties of the pseudo-inverse matrix:

$$(A^+)^+ = A, \tag{1.2.16}$$

$$(A^*)^+ = (A^+)^*. \tag{1.2.17}$$

Property (1.2.16) follows directly from the relationships (1.2.1), which define the pseudo-inverse matrix: the matrices A^+ and A enter into these relationships symmetrically.

In order to prove property (1.2.17), it is easiest to use Theorem 1.2.1. The point here is that if $A = B \cdot C$ is a skeleton expansion of the matrix A, then $A^* = C^*B^*$ is one of the skeleton expansions of the matrix A^*. But then, by replacing B with C^* and C with B^* in (1.2.2), we obtain:

$$(A^*)^+ = B(B^*B)^{-1}(CC^*)^{-1}C = [C^*(CC^*)^{-1}(B^*B)^{-1}B^*]^* = (A^+)^*.$$

From property (1.2.17) it follows that, if the matrix A is self-conjugate, i.e. $A^* = A$, then the pseudo-inverse matrix A^+ is also self-conjugate because in this case (by virtue of (1.2.17))

$$A^+ = (A^+)^*. \tag{1.2.18}$$

It is interesting to note that, for the self-conjugate matrix A, the system (1.2.1) is equivalent to another, simpler system:

$$AA^+ = A^+A, \qquad A^+ = A^+AA^+, \qquad A^+A^2 = A. \tag{1.2.19}$$

The proof of this fact is left to the reader.

The following theorem is important for our purposes.

Theorem 1.2.3 (see, for example, [3] as well as [2, p. 43])

$$A^+ = \lim_{\delta \to 0} (\delta^2 E + A^*A)^{-1} A^* = \lim_{\delta \to 0} A^*(\delta^2 E + AA^*)^{-1} \qquad (1.2.20)$$

and the estimator

$$\| A^+ - (\delta^2 E + A^*A)^{-1} A^* \| \leqslant \delta^2 \| A^+(A^+)^* A^+ \| \qquad (1.2.21)$$

holds (here $\| \cdot \|$ is a spectral norm, i.e. a norm obeying the Euclidean norm of vectors; the variable δ is real valued).

Proof Since $A^+ = A^+ AA^+$ and $(A^+A)^* = A^+A$, then $(A^+)^* = (A^+)^*A^+A$ and, therefore,

$$A^+ = A^*(A^+)^* A^+. \qquad (1.2.22)$$

Moreover, since $AA^+A = A$ and $(AA^+)^* = AA^+$, we have $[A^*(AA^+ - E)]^* = (AA^+ - E)A = 0$ and, therefore,

$$A^* = A^*AA^+. \qquad (1.2.23)$$

By substituting (1.2.23) into (1.2.22), we obtain:

$$A^+ = A^*AA^+(A^+)^* A^+. \qquad (1.2.24)$$

But then, in view of (1.2.23) and (1.2.24),

$$\begin{aligned}
A^+ - (\delta^2 E + A^*A)^{-1} A^* &= (\delta^2 E + A^*A)^{-1}[\delta^2 A^+ + A^*AA^+ - A^*] \\
&= \delta^2(\delta^2 E + A^*A)^{-1} A^*AA^+(A^+)^* A^+,
\end{aligned}$$

which, because

$$\| (\delta^2 E + A^*A)^{-1} A^*A \| \leqslant 1,$$

yields the estimator (1.2.21) and, therefore, the first equality of (1.2.20). The second equality of (1.2.20) is also valid because it is easy to ensure that $(\delta^2 E + A^*A)^{-1} A^* = A^*(\delta^2 E + AA^*)^{-1}$.

1.3. Semi-inverse matrices and matrix equations

Semi-inverse matrices can be used to concisely represent the simultaneity conditions and a general solution of the matrix equation

$$AXB = C, \qquad (1.3.1)$$

is the following.

Lemma 1.3.1 In order for equation (1.3.1) to be solvable (for X), it is necessary and sufficient that the following equalities hold:

$$(E - AA^-)C = 0, \qquad C(E - B^-B) = 0. \tag{1.3.2}$$

Necessity   Let the matrix X_1 be the solution of equation (1.3.1). Then, by multiplying the equality $AX_1B = C$ on the left by the matrix $E - AA^-$, by taking into account in this case property (1.1.2), we obtain the first of the equalities (1.3.2). Similarly, by multiplying the equality $AX_1B = C$ on the right by the matrix $E - B^-B$, we obtain the second of the equalities (1.3.2).

Sufficiency   If the equalities (1.3.2) are valid, then by means of direct substitution one can make sure that $X_1 = A^-CB^-$ is the solution of equation (1.3.1).

Lemma 1.3.2   If the matrix equation

$$AXB = C \tag{1.3.3}$$

is solvable, then its general solution is representable as

$$X = A^-CB^- + (E - A^-A)U + V(E - BB^-), \tag{1.3.4}$$

where A^- and B^- are any semi-inverse matrices for the matrices A and B, and U and V are arbitrary matrices of suitable dimensions.

Proof   In view of the solvability conditions (1.3.2) and of the equalities $AA^-A = A$ and $BB^-B = B$, it is easy to demonstrate by means of direct substitution that a matrix of the form (1.3.4) satisfies equation (1.3.3). At the same time any matrix X, which is the solution of equation (1.3.3), is representable as (1.3.4) when $U = X$ and $V = A^-AX - A^-CB^-$. This completes the proof of Lemma 1.3.2.

For a particular case of the algebraic system $Ax = f$, when the matrix B in equation (1.3.3) is a one-element one ($B = (1)$, $B^- = (1)$), and x and f are vectors, Lemmas 1.3.1 and 1.3.2 yield the corollary.

Corollary 1.3.1   The linear algebraic system

$$Ax = f \tag{1.3.5}$$

is solvable if and only if

$$(E - AA^-)f = 0, \tag{1.3.6}$$

and if it is solvable, then its general solution is representable as

$$x = A^-f + (E - A^-A)u, \tag{1.3.7}$$

where A^- is an arbitrary semi-inverse matrix for the matrix A, and u is an arbitrary vector.

In what follows the case is important when matrix elements of the algebraic system are certain functions. For this case, we shall give the following three lemmas.

Lemma 1.3.3 Let $\mathscr{M}(t) \in C[\alpha, \beta]$. Then any solution V of the system

$$\left[\int_\alpha^\beta \mathscr{M}^*(s)\mathscr{M}(s)\,ds \right] V = 0 \tag{1.3.8}$$

is a constant solution of the system

$$\mathscr{M}(t)V = 0, \qquad \alpha \leqslant t \leqslant \beta. \tag{1.3.9}$$

Conversely, any constant solution of the system (1.3.9) is the solution of the system (1.3.8).

Proof Note first that, for any matrix A, $A = 0$ follows from A^*A. Indeed, for any vector x, we have

$$(A^*Ax, x) = (Ax, Ax) = \| Ax \|^2,$$

and, therefore, if $A^*A = 0$, then $\| Ax \| = 0$ and, consequently, at any vector x, the equality $Ax = 0$ holds. But then $A = 0$.

Now let V be the solution of equation (1.3.8). By multiplying this equation by the matrix V^* on the left, we obtain:

$$\int_\alpha^\beta V^* \mathscr{M}^*(s)\mathscr{M}(s)V\,ds = 0, \tag{1.3.10}$$

and, since the integrand function in (1.3.10) is continuous, from (1.3.10) it follows that it identically equals zero. But then, by virtue of the above remark, $\mathscr{M}(t)V = 0$, for all $t \in [\alpha, \beta]$. The second assertion of the lemma is obvious.

Lemma 1.3.4 In the system let

$$\mathscr{M}(t)z = \mathscr{N}(t), \qquad \alpha \leqslant t \leqslant \beta, \qquad \mathscr{M}(t) \in C[\alpha, \beta], \qquad \mathscr{N}(t) \in C[\alpha, \beta]. \tag{1.3.11}$$

Then the system (1.3.11) has the $t \in [\alpha, \beta]$ independent solution z if and only if at all $t \in [\alpha, \beta]$

$$\mathscr{N}(t) = \mathscr{M}(t) \left[\int_\alpha^\beta \mathscr{M}^*(s)\mathscr{M}(s)\,ds \right]^- \int_\alpha^\beta \mathscr{M}^*(s)\mathscr{N}(s)\,ds. \tag{1.3.12}$$

Proof Let z be a constant solution of the system (1.3.11). Then, obviously, the equality

$$\int_\alpha^\beta \mathscr{M}^*(s)\mathscr{M}(s)\,ds \cdot z = \int_\alpha^\beta \mathscr{M}^*(s)\mathscr{N}(s)\,ds \tag{1.3.13}$$

is satisfied.

The system (1.3.13) is a system with constant matrices and, therefore, by virtue of Corollary 1.3.1,

$$z = \left[\int_\alpha^\beta \mathscr{M}^*(s)\mathscr{M}(s)\,ds \right]^- \int_\alpha^\beta \mathscr{M}^*(s)\mathscr{N}(s)\,ds + V, \qquad (1.3.14)$$

with a certain matrix V satisfying the equation

$$\int_\alpha^\beta \mathscr{M}^*(s)\mathscr{M}(s)\,ds \cdot V = 0. \qquad (1.3.15)$$

But then, upon substitution of (1.3.14) into (1.3.11) (subject to Lemma 1.3.3), one obtains the equality (1.3.12).

If, however, equality (1.3.12) is satisfied, then obviously a constant solution of the system (1.3.11) exists (it is the product of the two last cofactors in (1.3.12)).

Lemma 1.3.5   Under the suppositions of Lemma 1.3.4 a general constant solution of the system (1.3.11) is representable as (1.3.14), where V is a general constant solution of the system (1.3.15).

Proof   By virtue of the solvability condition (1.3.12), the first summand in (1.3.14) is a partial (constant) solution of the system (1.3.11), and the second summand is, by virtue of Lemma 1.3.3, a general (independent of t) solution of the homogeneous system $\mathscr{M}(t)V = 0$. This completes the proof of Lemma 1.3.5.

Note that, with respect to the algebraic system

$$Ax = f \qquad (1.3.16)$$

(with constant matrix A and constant vector f), the following assertions follow from Lemmas 1.3.4 and 1.3.5.

Corollary 1.3.2   The system (1.3.16) is joint if and only if

$$f = A(A^*A)^- A^* f. \qquad (1.3.17)$$

Corollary 1.3.3   If the system (1.3.16) is joint, then its general solution is representable as

$$x = (A^*A)^- A^* f + v, \qquad (1.3.18)$$

where v is an arbitrary solution of the homogeneous system $Av = 0$.

Corollaries 1.3.2 and 1.3.3 lead one to suggest that the matrix $(A^*A)^- A^*$ is a semi-inverse matrix for the matrix A, and this is, indeed, confirmed by the following lemma.

Lemma 1.3.6   The matrices

$$(A^*A)^- A^*, \qquad A^*(AA^*)^-$$

are semi-inverse for the matrix A.

Proof Let us prove first that, with arbitrary matrices A and B, $AB = 0$ follows from $A^*AB = 0$.

Indeed, we note that, if $A^*AB = 0$, then with an arbitrary vector x

$$(A^*ABx, Bx) = 0.$$

But then

$$\|ABx\|^2 = (ABx, ABx) = (A^*ABx, Bx) = 0$$

and, consequently, at an arbitrary vector x, the equality $ABx = 0$ holds and, therefore, the equality $AB = 0$ holds too.

Note now that, by virtue of the determining property of the semi-inverse matrices (see (1.1.1)), $A^*A[(A^*A)^- A^*A - E] = 0$. But then, according to what we have just proved,

$$A[(A^*A)^- A^*A - E] = A(A^*A)^- A^*A - A = 0$$

and, consequently, the matrix $(A^*A)^- A^*$ is indeed semi-inverse to the matrix A.

Proceeding in just the same way, one can make sure that the matrix $(AA^*)^- A$ is a semi-inverse matrix for the matrix A^*, and this, subject to Lemma 1.1.8 and the equalities (1.1.16), leads to the following equality:

$$A^*(AA^*)^- = [(AA^*)^- A]^* = [(A^*)^-]^* = A^-.$$

Lemma 6 is thereby proved.

1.4. Projectors and mappings

The results presented in the preceding sections permit us to construct some important projectors and mappings of linear vector spaces. The following lemmas, for example, are valid.

Lemma 1.4.1 Let the matrices A and B have dimensions $(m_1 \times n)$ and $(m_2 \times n)$, respectively. Then[†]

$$\ker A \cap \ker B = \operatorname{Im} P, \tag{1.4.1}$$

where

$$P = (E - A^- A)[E - (B - BA^- A)^- (B - BA^- A)], \tag{1.4.2}$$

and the matrix P is a projector of space R_n onto the intersection of the kernels of the matrices A and B.

[†]From here on we shall be using the following designations: $\ker A$ = matrix A kernel; $\operatorname{Im} A$ = image of matrix A; $\dim L$ = dimensions of space L; $\operatorname{def} A = \dim (\ker A)$; and rank = $\dim (\operatorname{Im} A)$.

Proof If $x \in \ker A \cap \ker B$, then x satisfies the system

$$Ax = 0, \qquad Bx = 0 \tag{1.4.3}$$

and, while satisfying the first equation of this system, by virtue of Corollary 1.3.1, with a certain vector u, is representable as

$$x = (E - A^- A)u. \tag{1.4.4}$$

On substituting (1.4.4) into the second equation of (1.4.3), we find that the vector u satisfies the equality $(B - BA^- A)u = 0$. But then (again by virtue of Corollary 1.3.1) the representation

$$u = [E - (B - BA^- A)^- (B - BA^- A)]v, \tag{1.4.5}$$

holds for the vector u at a certain v. We now substitute (1.4.5) into (1.4.4) and, subject to the notation (1.4.2), we obtain $x = Pv$, i.e. $x \in \text{Im } P$. Hence, the left-hand side of (1.4.1) belongs to the right-hand side. In order to prove the inverse inclusion, we note first that $AP = 0$ and $BP = 0$, which is a simple consequence of the determining property (1.1.1) of semi-inverse matrices. And if now $x \in \text{Im } P$, i.e. the equality $x = Pv$ is valid for a certain vector v, then $Ax = 0$ and $Bx = 0$ and, consequently, $x \in \ker A \cap \ker B$. The first assertion of Lemma 1.4.1 is proved.

In order to prove that P is a projector, it still remains to demonstrate that, if $x \in \ker A \cap \ker B$, then $Px = x$. But this is simply established by means of the determining property (1.1.1) of semi-inverse matrices. The lemma is proved.

Corollary 1.4.1 For the $(m \times n)$ matrix A,

$$\ker A = \text{Im}(E - A^- A)$$

and the matrix $E - A^- A$ is a projector of the space R_n onto the nucleus of the matrix A.

In order to prove Corollary 1.4.1, it is sufficient in the formulation of Lemma 1.4.1 to put $B = 0$.

Corollary 1.4.2 With respect to the block matrix $\begin{pmatrix} A \\ B \end{pmatrix}$ the relationship

$$\text{def}\begin{pmatrix} A \\ B \end{pmatrix} = \text{rank}\{(E - A^- A)[E - (B - BA^- A)^- (B - BA^- A)]\} \tag{1.4.6}$$

holds. In particular, when $B = 0$

$$\text{def } A = \text{rank}(E - A^- A). \tag{1.4.7}$$

Proof Since, obviously,

$$\ker\begin{pmatrix} A \\ B \end{pmatrix} = \ker A \cap \ker B,$$

then Corollary 1.4.2 follows immediately from equality (1.4.1) and from the definitions of rank and defect: the rank of matrix $\mathscr{M}$ is the dimensions of its image, and the defect is the dimensions of its nucleus.

Lemma 1.4.2 Let the matrices $A_1, A_2, \ldots, A_N$ have the dimensions $(m_i \times n)$ $(i = 1, 2, \ldots, N)$, respectively, and let the matrix P_{N-1} be a projector of the space R_n onto the intersection of the nuclei of the matrices $A_1, \ldots, A_{N-1}$. Then the matrix

$$P_N = P_{N-1}[E - (A_N P_{N-1})^- (A_N P_{N-1})] \tag{1.4.8}$$

is a projector onto the intersection of the nuclei of the matrices $A_1, \ldots, A_{N-1}, A_N$.

Proof Since, by virtue of the assumption, for any vector $x A_i P_{N-1} x = 0$ $(i = 1, \ldots, N - 1)$, then $A_i P_{N-1} = 0$ and, consequently, in view of (1.4.8), $A_i P_N = 0$ $(i = 1, \ldots, N - 1)$. Besides, from (1.4.8) and from the determinant property of semi-inverse matrices it follows that $A_N P_N^- = 0$. Therefore, for any vector $x \in R_n$, we have $A_i P_N x = 0$ $(i = 1, \ldots, N)$, i.e. $\operatorname{Im} P_N \in \bigcap_{i=1}^N \ker A_i$. It remains to make sure that $P_N x = x$ follows from $x \in \bigcap_{i=1}^N \ker A_i$. But if $x \in \bigcap_{i=1}^N A_i$, then $x \in \bigcap_{i=1}^{N-1} \ker A_i$, $x \in \ker A_N$ and, therefore (in view of the conditions of the lemma), $P_{N-1} x = x$, $A_N x = 0$ and, by virtue of formula (1.4.8), $P_N x = x$. The lemma is proved.

If Corollary 1.4.1 is taken into account and the matrix

$$P_1 = E - A_1^- A_1$$

is taken as the matrix P_1, then by formula (1.4.8) it is possible to recurrently construct the projectors, first onto the intersection of the nuclei of the matrices A_1, A_2, then onto the intersection of the nuclei of the matrices A_1, A_2, A_3, etc. This makes it possible to apply formulae (1.4.8) to solve the homogeneous system $Ax = 0$ with the matrix

$$A = \begin{pmatrix} A_1 \\ A_2 \\ \vdots \\ A_N \end{pmatrix} \tag{1.4.9}$$

(since the problem of solving such a system is equivalent to the problem of determining the projector onto the intersection of the nuclei of the matrices A_1, $A_2, \ldots, A_N$). In particular, the rows of the matrix A may be taken as the matrices $A_i (i = 1, \ldots, N)$ in (1.4.9), and then formula (1.4.8) will contain the semi-inverse only to one-row matrices.

Lemma 1.4.3 Let the matrices A and B have the dimensions $(m \times n_1)$ and

$(m \times n_2)$, respectively. Then

$$\operatorname{Im} A + \operatorname{Im} B = \ker Q, \qquad (1.4.10)$$

where $Q = [E - (B - AA^-B)(B - AA^-B)^-](E - AA^-)$ and, consequently, the block matrix (A, B) maps the space $R_{n_1 + n_2}$ to the nucleus of the matrix Q.

Proof If x belongs to the left-hand side of (1.4.10), then

$$x = Au + Bv \qquad (1.4.11)$$

for certain vectors $u \in R_{n_1}$, $v \in R_{n_2}$. But from the determining property of semi-inverse matrices (1.1.1) it readily follows that $QA = 0$ and $QB = 0$ and, therefore, $Qx = 0$ and $x \in \ker Q$, i.e. the vector x also belongs to the right-hand side of (1.4.10). Now let $x \in \ker Q$ and, consequently, $Qx = 0$. Then, by virtue of Corollary 1.3.1, the system

$$(B - AA^-B)y = (E - AA^-)x$$

is solvable for y and, consequently, as is easy to verify, the vector x is representable as (1.4.11) when

$$u = A^-(x - By), \qquad v = y.$$

But this does indeed mean that x belongs to the left-hand side of (1.4.10). The equality (1.4.10) is proved. The second assertion of the lemma is now obvious.

Corollary 1.4.3 For the $(m \times n)$ matrix A

$$\operatorname{Im} A = \ker(E - AA^-)$$

and, consequently, the matrix A maps the space R_n onto the nucleus of the matrix $E - AA^-$.

For the proof, it is necessary, first, to put $B = 0$ in Lemma 1.4.3.

Corollary 1.4.4 For the block matrix (A, B), the relationship

$$\operatorname{rank}(A, B) = \operatorname{def}\{[E - (B - AA^-B)(B - AA^-B)^-](E - AA^-)\}$$

holds. In particular, when $B = 0$ rank $A = \operatorname{def}(E - AA^-)$.

Proof Since $\operatorname{Im}(A, B) = \operatorname{Im} A + \operatorname{Im} B$, then Corollary 1.4.4 follows from formula (1.4.10) and from the definitions of rank and defect.

Lemma 1.4.4 Let the matrices A and B have the dimensions $(m \times n_1)$ and $(m \times n_2)$, respectively. Then

$$\operatorname{Im} A \cap \operatorname{Im} B = \operatorname{Im} P = \operatorname{Im} Q, \qquad (1.4.12)$$

where

$$P = B[E - (B - AA^-B)^-(B - AA^-B)],$$
$$Q = A[E - (A - BB^-A)^-(A - BB^-A)],$$

and the matrix P maps the space R_{n_2}, and the matrix Q maps the space R_{n_1} onto the intersection of the mappings of the matrices A and B.

Proof If $x \in \operatorname{Im} A \cap \operatorname{Im} B$, then x has two representations:

$$x = Au, \qquad x = Bv,$$

where u and v are related by the equality $Au = Bv$. By multiplying this equality by the matrix $E - AA^-$, for the vector v, we obtain $(B - AA^- B)v = 0$. Therefore, according to Corollary 1.3.1, v is representable as

$$v = [E - (B - AA^- B)^-(B - AA^- B)]w,$$

and because $x = Bv$, then $x = Pw$, and this means that $x \in \operatorname{Im} P$.

If, however, $x \in \operatorname{Im} P$, i.e. x has the form $x = Pw$, then since in the expression for P the first co-factor is the matrix B, we have $x \in \operatorname{Im} B$. At the same time, as is easy to verify, $(E - AA^-)P = 0$, and therefore $x = AA^- P$ and, consequently, $x \in \operatorname{Im} A$. Thus, x belongs to the left-hand side of the equality (1.4.12) and, consequently, the first assertion of Lemma 1.4.4 with respect to the matrix P is proved. The proof of the first assertion with respect to the matrix Q is carried out in a similar way. The second assertion of Lemma 1.4.4 becomes obvious after that.

Corollary 1.4.5 For any $(m \times n)$ matrix A

$$\operatorname{Im} A = \operatorname{Im}(AA^-), \tag{1.4.13}$$

and the matrix AA^- maps the space R_m onto the mapping of the matrix A.

For the proof, it is necessary to put $B = E$ in the formulation of Lemma 1.4.4 and to use the determining property (1.1.1) of semi-inverse matrices.

Note that from the equality (1.4.13) it follows that

$$\operatorname{rank} A = \operatorname{rank} AA^-. \tag{1.4.14}$$

Lemma 1.4.5 Let the matrices A and B have the dimensions $(m_1 \times n)$ and $(m_2 \times n)$, respectively. Then

$$\ker A + \ker B = \ker P, \tag{1.4.15}$$

where $P = [E - (B - BA^- A)(B - BA^- A)^-]B$.

Proof Let $x \in \ker A + \ker B$, i.e. $x = u + v$, where $Au = 0$ and $Bv = 0$. Then, as a consequence of the equality $P(E - A^- A) = 0$, we have

$$Px = Pu + Pv = PA^- Au + Pv = Pv = [E - (B - BA^- A)(B - BA^- A)^-]Bv = 0,$$

and, therefore, $x \in \ker P$. If, however, $x \in \ker P$, i.e. $Px = 0$, then by virtue of Corollary 1.3.1, the equation $(B - BA^- A)y = Bx$ is solvable for y. But then, if u is the solution of this equation, then (again by virtue of the same corollary), there exists a vector v such that

$$(E - A^- A)u = x - (E - B^- B)v.$$

Hence, $x = u_1 + v_1$, where, as is easy to see,

$$u_1 = [E - A^- A]u \in \ker A, \qquad v_1 = (E - B^- B)v \in \ker B.$$

Consequently, $x \in \ker A + \ker B$—the left-hand side of the equality (1.4.15). The lemma is proved.

Corollary 1.4.6   For any $(m \times n)$ matrix A

$$\ker A = \ker A^- A$$

and, consequently,

$$\operatorname{def} A = \operatorname{def}(A^- A).$$

For the proof, it is necessary to put $B = E$ in Lemma 1.4.5 and to take into consideration that this matrix itself is a semi-inverse matrix to the matrix $(E - A^- A)$.

Lemma 1.4.6   Let the matrices A and B have the dimensions $(m \times n)$ and $(m_1 \times m)$, respectively. Then

$$\operatorname{Im} A \cap \ker B = \operatorname{Im} P, \qquad (1.4.16)$$

where $P = A[E - (BA)^- (BA)]$. In this case the matrix P performs the mapping of the space R_n onto the intersection $\operatorname{Im} A \cap \ker B$.

Proof   Let x belong to the left-hand side of the equality (1.4.16). Then $Bx = 0$ and, with a certain vector u, $x = Au$. In this case the vector u, of course, satisfies the equality $BAu = 0$. This equality, according to Corollary 1.3.1 for the vector u, yields the representation

$$u = [E - (BA)^- BA]v,$$

and since $x = Au$, then $x = Pv$, i.e. the vector x belongs to the right-hand side of the equality (1.4.16).

Conversely, if $x \in \operatorname{Im} P$, then obviously $x \in \operatorname{Im} A$, because the first factor in the formula for P is the matrix A. Moreover, since $BP = 0$, then $x \in \ker B$ and, consequently, x belongs to the left-hand side of the equality (1.4.16). The first assertion of the lemma is proved. The second assertion is now obvious.

Lemma 1.4.7   Let the matrices A and B have the dimensions $(m \times n)$ and $(n \times n_1)$, respectively. Then

$$\ker A + \operatorname{Im} B = \ker P, \qquad (1.4.17)$$

where

$$P = [E - (AB)(AB)^-]A.$$

Proof If $x \in \ker A + \operatorname{Im} B$, then $x = u + v$, where $Au = 0$ and the vector v, with a certain w, is representable as $v = Bw$. By multiplying the equality $x = u + Bw$ by the matrix A, we obtain $Ax = ABw$. Thus, $x \in \ker P$. Conversely, if $x \in \ker P$, i.e. $Px = 0$, then by virtue of Corollary 1.3.1, the equation $ABy = Ax$ is solvable for y and there exists such a vector u that $By = x + (E - A^{-}A)u$ and, therefore, $x = u_1 + u_2$, where $u_1 = By \in \operatorname{Im} B$, $u_2 = -(E - A^{-}A)u \in \ker A$, i.e. x belongs to the left-hand side of the equality (1.4.17). The lemma is proved.

The following lemmas give expressions for ranks of block matrices and products of matrices.

Lemma 1.4.8 For the block matrix (A, B)

$$\operatorname{rank}(A, B) = \operatorname{rank} A + \operatorname{rank} B - \operatorname{rank}\{B[E - (B - AA^{-}B)^{-}(B - AA^{-}B)]\}.$$

Proof With respect to the two subspaces L_1 and L_2 of a linear vector space R_m, we have the equality

$$\dim(L_1 + L_2) = \dim L_1 + \dim L_2 - \dim(L_1 \cap L_2)$$

(Grassman's formula; see [4, Problem 60]). But then

$$
\begin{aligned}
\operatorname{rank}(A, B) &= \dim(\operatorname{Im}(A, B)) = \dim(\operatorname{Im} A + \operatorname{Im} B) \\
&= \dim(\operatorname{Im} A) + \dim(\operatorname{Im} B) - \dim(\operatorname{Im} A \cap \operatorname{Im} B) \\
&= \operatorname{rank} A + \operatorname{rank} B - \dim(\operatorname{Im} A \cap \operatorname{Im} B),
\end{aligned}
$$

and since, by virtue of Lemma 1.4.4

$$
\begin{aligned}
\dim(\operatorname{Im} A \cap \operatorname{Im} B) &= \dim(\operatorname{Im} P) = \operatorname{rank} P \\
&= \operatorname{rank}\{B[E - (B - AA^{-}B)^{-}(B - AA^{-}B)]\},
\end{aligned}
$$

then Lemma 1.4.8 is proved.

Lemma 1.4.9 Let the matrices A and B have the dimensions $(m \times n)$ and $(m_1 \times m)$, respectively. Then

$$\operatorname{rank}(BA) = \operatorname{rank} A - \operatorname{rank}\{A[E - (BA)^{-}(BA)]\}. \tag{1.4.18}$$

Proof This lemma is a simple corollary of the known equality

$$\operatorname{rank}(BA) = \operatorname{rank} A - \dim(\operatorname{Im} A \cap \ker B)$$

(see [4, Problem 133]) and Lemma 1.4.6.

Corollary 1.4.7 For the $(m \times n)$ matrix A we have the equality

$$\operatorname{rank} A = n - \operatorname{rank}(E - A^{-}A). \tag{1.4.19}$$

In order to prove this assertion, it is sufficient to substitute the matrix A

instead of the matrix B into the equality (1.4.18), and instead of the matrix A, the $(n \times n)$ matrix E. As a result, we obtain the equality (1.4.19).

Corollary 1.4.8   The equality

$$\text{rank}(A^- A) = \text{rank } A$$

holds (cf. (1.4.14)).

In order to check this, it is sufficient to substitute the matrix A^- instead of the matrix B into the equality (1.4.18). In this case it is necessary to take into consideration that one of the semi-inverse matrices for the matrix $A^- A$ is the matrix itself and $(A^- A)^2 = A^- A$, as a consequence of which $A[E - (A^- A)^-(A^- A)] = 0$.

Lemma 1.4.10   Let the matrices A and B have the dimensions $(m \times n)$ and $(n \times n_1)$, respectively. Then

$$\text{rank}(AB) = \text{rank } A - \text{rank}\{[E - (AB)(AB)^-]A\}. \tag{1.4.20}$$

Proof   It is easiest to prove this lemma by applying Lemmas 1.1.8 and 1.4.9 to the matrix B^*A^* and conjugating to the matrix AB. Namely, by virtue of Lemma 1.4.9, $\text{rank}(B^*A^*) = \text{rank } A^* - \text{rank}\{A^*[E - (B^*A^*)^-(B^*A^*)]\}$,   and since the rank of any matrix is equal to the rank of the matrix conjugating to it, and by Lemma 1.1.8

$$[(B^*A^*)^-]^* = (AB)^-,$$

we have the equality (1.4.20), as was to be shown.

Corollary 1.4.9   For the $(m \times n)$ matrix A

$$\text{rank } A = m - \text{rank}(E - AA^-).$$

In order to prove Corollary 1.4.8, it is sufficient to substitute the matrix A instead of the matrix B into the equality (1.4.20), and instead of the matrix A, the $(m \times m)$ matrix E.

Another proof of the assertions of Lemmas 1.4.9 and 1.4.10 is reported in [5, p. 154].

Next, by taking into account the equalities (1.4.7) and (1.4.19), for the $(m \times n)$ matrix A we obtain:

$$\text{rank } A + \text{def } A = n. \tag{1.4.21}$$

In what follows other forms of the formulations of Lemmas 1.4.9 and 1.4.10 will be useful for our purposes. We shall give them in the form of independent lemmas.

Lemma 1.4.11   Let the matrices A and B have the dimensions $(m \times n)$ and

$(m_1 \times m)$, respectively. Then

$$\text{def}(BA) = \text{def } A + \text{rank}\{A[E - (BA)^-(BA)]\}. \tag{1.4.22}$$

In order to prove the lemma, it is sufficient to extract the left- and right-hand sides of (1.4.18) from n and to apply the relationship (1.4.21).

Lemma 1.4.12 Let the matrices A and B have the dimensions $(m \times n)$ and $(n \times n_1)$, respectively. Then

$$\text{def}(AB) = \text{def } A + \text{rank}\{[E - (AB)(AB)^-]A\} + n_1 - n. \tag{1.4.23}$$

In order to prove this assertion, it is sufficient to extract the left- and right-hand sides of the equality (1.4.20) from n_i to add $n_1 - n$ to the right-hand side of the equality thus obtained, and to apply the relationship (1.4.21).

Corollary 1.4.10 For any $(m \times n)$ matrix the equality

$$\text{def}(AA^-) = \text{def } A + m - n$$

holds.

In order to prove Corollary 1.4.10, it is sufficient to put $B = A^-$ into the equality (1.4.23) and to take into consideration that $(AA^-)^- = AA^-$ and $(AA^-)^2 = AA^-$ and, therefore, that $[E - (AA^-)(AA^-)^-]A = 0$.

Note further that if the rank of the $(m \times n)$ matrix A is equal to n, then, by virtue of the equality (1.4.19), $\text{rank}(E - A^-A) = 0$ and, consequently, $E - A^-A = 0$. If, however, $\text{rank } A = m$, then according to Corollary 1.4.9 $(E - AA^-) = 0$ and, therefore, $E - AA^- = 0$. Inverse assertions also seem to be valid, and since the rank of the matrix is equal to the number of its linearly independent rows or columns, we have a further two lemmas.

Lemma 1.4.13 The equality $E - A^-A = 0$ is valid if and only if the columns of the $(m \times n)$ matrix A are linearly independent $(\text{rank } A = n)$.

Lemma 1.4.14 The equality $E - A^-A = 0$ holds if and only if the rows of the $(m \times n)$ matrix are linearly independent $(\text{rank } A = m)$.

Further useful information on generalized inverse matrices may be derived from [3, 6].

1.5. Quite perfect pairs of matrices

When constructing algorithms for solving degenerate systems of the form $Ax' = Bx + f$, there primarily arise the following questions.

(1) Is it possible to reduce the solution of the algebraic system

$$A(\lambda x) = Bx + f \qquad (1.5.1)$$

to the solution of the system with the matrix $\lambda E - A^- B$?
(2) Does there exist a number λ for which

$$\det(\lambda A - B) = 0? \qquad (1.5.2)$$

It is especially easy to obtain answers to these questions in the case when the pair of matrices (A, B) is the so-called quite perfect pair.

Definition 1.5.1   The pair of matrices (A, B) is said to be quite perfect if

$$(E - AA^-)B = 0 \qquad (1.5.3)$$

(from here on, the pair of matrices (A, B) is assumed to be an ordered one, i.e. $(A, B) \neq (B, A)$).
   If a pair of matrices in the system (1.5.1) is quite perfect, then, using Corollary 1.3.1, it is easy to show that the solution of the system (1.5.1) is reduced to the solution of the system

$$\lambda x = A^- Bx + A^- f + u, \qquad (1.5.4)$$

$$Au = 0, \qquad (1.5.5)$$

and the equality

$$(E - AA^-)f = 0 \qquad (1.5.6)$$

is the compatibility condition for the system (1.5.1).
   Thus, the answer to the first question with respect to the system (1.5.1), the pair of matrices of which is quite perfect, is obtained.
   Suppose now that λ is not an eigen-number of the matrix $A^- B$. With such a supposition, equation (1.5.4) is uniquely solvable for x, and since the solution of equation (1.5.5) has the form $u = (E - A^- A)v$, then for a general solution of equation (1.5.1) we obtain the formula

$$x = (\lambda E - A^- B)^{-1}[A^- f + (E - A^- A)v], \qquad (1.5.7)$$

where v is an arbitrary vector. The equality (1.5.6) in this case becomes a necessary and sufficient condition of solvability.
   In the case of a homogeneous system

$$A(\lambda x) = Bx \qquad (1.5.8)$$

formula (1.5.7) takes on the form

$$x = (\lambda E - A^- B)^{-1}(E - A^- A)v,$$

from which, as a consequence of the arbitrariness of v, it follows that equation (1.5.8) has only a zero solution (or, equivalently, that equality (1.5.2) is valid) if and only if $E - A^- A = 0$ and since the equality $E - A^- A = 0$, by virtue of Lemma 1.4.13 is equivalent to a linear independence of the matrix A columns, such an assertion is proved.

Lemma 1.5.1 Let λ not be an eigen-number of the matrix $A^- B$ and the pair of matrices (A, B) be quite perfect. Then the equality (1.5.2) is valid if and only if the columns of the matrix A are linearly independent.

When the lemma is proved it gives the answer to the second question in the case of a quite perfect pair of matrices.

When considering an arbitrary pair of matrices (A, B), the notion of a quite perfect pair of matrices is also useful: by applying this notion, it becomes possible to reduce the solution of an arbitrary system (1.5.1) to the solution of a certain system with a perfect pair of matrices and thereby to obtain answers to the questions posed in the general case.

In order to understand the reason behind this, we first consider two chains of matrices,

$$A_0 = A, A_1, A_2, \ldots, A_i, \ldots, \tag{1.5.9}$$

$$B_0 = B, B_1, B_2, \ldots, B_i, \ldots, \tag{1.5.10}$$

which are obtained with the help of the following recurrent formulae:

$$\begin{aligned} A_{i+1} &= A[E - (B - A_i A_i^- B)^- (B - A_i A_i^- B)], \\ B_{i+1} &= B[E - (B - A_i A_i^- B)^- (B - A_i A_i^- B)], \end{aligned} \quad i = 0, 1, \ldots . \quad \begin{aligned} &\text{(1.5.11)} \\ &\text{(1.5.12)} \end{aligned}$$

Note that, by virtue of Lemma 1.4.4, $\operatorname{Im} B_{i+1} = \operatorname{Im} A_i \cap \operatorname{Im} B$. However, we shall not need this: by using semi-inverse matrices it becomes possible to get rid of considerations of the various subspaces by reducing them to calculations. For example, the equality

$$B_{i+1} = A_i A_i^- B_{i+1}, \tag{1.5.13}$$

important for our purposes, follows immediately from the determining property of semi-inverse matrices.

Furthermore, it appears that, without changing the chain (1.5.9), formula (1.5.11) can be replaced by the formula

$$A_{i+1} = A_i[E - (B - A_i A_i^- B)^- (B - A_i A_i^- B)], \tag{1.5.14}$$

from which follows the equality

$$A_{i+1} = A_i A_i^- A_{i+1}, \tag{1.5.15}$$

which is similar to (1.5.13).

Before proving formula (1.5.14), for the sake of brevity of the notation, we introduce the following designations:

$$P_i = E - A_i A_i^-, \qquad T_i = P_i B, \qquad R_i = E - T_i^- T_i. \tag{1.5.16}$$

With these designations, formulae (1.5.11), (1.5.12), and (1.5.14) have, respectively, the form

$$A_{i+1} = AR_i, \qquad B_{i+1} = BR_i, \tag{1.5.17}$$

$$A_{i+1} = A_i R_i \tag{1.5.18}$$

and one has to prove that the first formula in (1.5.17) can be written as (1.5.18). The proof will be carried out by induction in i.

When $i = 0$, the assertion is true because $A_0 = A$. Let us make sure that, from the validity of this assertion at a certain $i \geqslant 0$, its validity when $i = 1$ follows.

Thus, let $A_{i+1} = A_i R_i$. By multiplying this equality by the matrix P_i (see (1.5.16)), we obtain $P_i A_{i+1} = 0$. But then, by virtue of (1.5.17), (1.5.16), and (1.5.13),

$$R_i R_{i+1} = R_{i+1} - T_i^- T_i R_{i+1} = R_{i+1} - T_i^- P_i B R_{i+1} = R_{i+1} - T_i^- P_i B_{i+2}$$
$$= R_{i+1} - T_i^- (P_i A_{i+1}) A_{i+1}^- B_{i+2} = R_{i+1}$$

and, therefore,

$$A_{i+2} = AR_{i+1} = AR_i R_{i+1} = A_{i+1} R_{i+1},$$

which was to be shown.

Now, we note that, since the matrix R_i in formula (1.5.18) is quadratic, according to Lemma 1.4.12

$$\operatorname{def} A_{i+1} \geqslant \operatorname{def} A_i, \tag{1.5.19}$$

and since the defect of the matrix does not exceed the number of its columns (see (1.4.21)), the inequalities (1.5.19) when $i \to \infty$ must, without fail, involve an equality. In this case the following lemma is valid.

Lemma 1.5.2   If r is an integer positive number, at which

$$\operatorname{def} A_{r-1} = \operatorname{def} A_r, \tag{1.5.20}$$

then $P_r B_r = 0$.

Proof   Since the matrix R_{r-i} in the equality $A_r = A_{r-1} R_{r-1}$ is quadratic, then Lemma 1.4.12 yields

$$\operatorname{def} A_r = \operatorname{def} A_{r-1} + \operatorname{rank}[(E - A_r A_r^-) A_{r-1}],$$

and, therefore, as a consequence of (1.5.20)

$$\operatorname{rank}[(E - A_r A_r^-) A_{r-1}] = 0,$$

i.e. $P_r A_{r-1} = 0$, from which, in view of (1.5.13), we have the equality $P_r B_r = P_r A_{r-1} A_{r-1}^- B_r = 0$, and this completes the proof of Lemma 1.5.2.

Lemma 1.5.3   If r is a non-negative integer number, at which $P_r B_r = 0$, then $\operatorname{def} A_r = \operatorname{def} A_{r+1}$.

Proof Let r be a number satisfying the condition of the lemma. Then, taking into account (1.5.16)–(1.5.18), one can write a chain of equalities

$$A_{r+1}R_{r-1} = A_r R_r R_{r-1} = AR_r R_{r-1} = AR_{r-1} - AT_r^- P_r B_r R_{r-1} = A_r$$

(when $r = 0$, one should adopt $R_{-1} = E$ here). From the obtained equality $A_r = A_{r+1} R_{r-1}$, in which the matrix R_{r-i} is quadratic, the inequality

$$\operatorname{def} A_r \geqslant \operatorname{def} A_{r+1}$$

follows. On the other hand (see (1.5.19)),

$$\operatorname{def} A_r \leqslant \operatorname{def} A_{r+1},$$

and since the validity of the two opposite inequalities implies an equality, Lemma 1.5.3 is proved.

Corollary 1.5.1 If $P_r B_r = 0$, then $P_i B_i = 0$ when $i > r$.

Proof If $P_r B_r = 0$, then according to Lemma 1.5.3, $\operatorname{def} A_r = \operatorname{def} A_{r+1}$. But then according to Lemma 1.5.2, $P_{r+1} B_{r+1} = 0$, etc.

In order to better understand the meaning of the numbers r involved in Lemmas 1.5.2 and 1.5.3, we shall prove the independence of the defects of the matrices A_i of the particular choice of the semi-inverse matrices applied when constructing the chains (1.5.17). To do so, we consider the system

$$
\begin{aligned}
Bx_1 &= Ax_2, \\
Bx_2 &= Ax_3, \\
&\cdots \\
Bx_{i-1} &= Ax_i, \\
Bx_i &= Ax_{i+1}.
\end{aligned}
\tag{1.5.21}
$$

The last equation of this system is solvable for x_{i+1} if and only if x_i satisfies the equation $(E - AA^-)Bx_i \equiv T_0 x_i = 0$ (see Corollary 1.3.1). Consequently, the vector x_i, satisfying the system (1.5.21), will be represented as $x_i = R_0 y_i$ (see designation (1.5.16) and Corollary 1.3.1). By substituting this expression into the penultimate equation of the system (1.5.21), we obtain $B x_{i-1} = A_1 y_i$. Furthermore, as above, we make sure that the obtained equation is solvable for y_i if and only if the vector x_{i-1} has the form $x_{i-1} = R_1 y_{i-1}$.

Similar conclusions may be drawn also with regard to the other unknowns in the system (1.5.21) and, therefore, the general solution of this system is written as

$$x_j = R_{i-j} y_j, \qquad R_{-1} = E, \quad j = 1, \ldots, i+1, \tag{1.5.22}$$

where $y_j (j = 1, \ldots, i+1)$ is a general solution of the system

$$B_{i-j+1} y_j = A_{i-j} y_{j+1}, \quad j = 1, \ldots, i+1. \tag{1.5.23}$$

In order to obtain a general solution of the system (1.5.23), we note that each of the equations (1.5.23) is solvable for y_{j+1} at any y_j because, with any vector y_j, the corresponding compatibility condition $P_{i-j}B_{i-j+1}y_j = 0$, by virtue of (1.5.13), is satisfied. But then

$$y_{j+1} = A_{i-j}^- B_{i-j+1}y_j + z_j, \qquad (1.5.24)$$

where z_j is an arbitrary solution of the equation $A_{i-j}z_j = 0$.

Hence, $y_2 = A_{i-1}^- B_i y_1 + z_1$, where y_i is an arbitrary vector. The other unknowns in the system (1.5.23) are determined by the recurrent relationship (1.5.24). If the obtained values for y_j are now substituted into (1.5.22), then we find a general solution of the system (1.5.21). In this case, in particular, it will appear that $x_1 = R_{i-1}y_1$, where y_i is an arbitrary vector, i.e. the set $\mathcal{M}$ of values of x_i in the general solution of the system (1.5.21) is the map of the matrix $R_{i-1} : \mathcal{M} = \operatorname{Im} R_{i-1}$.

Obviously $\mathcal{M}$ is independent of any choice of the semi-inverse matrices because the system (1.5.21) does not involve any semi-inverse matrices at all. But then the set $A \cdot \mathcal{M}$, which is the map of the matrix $A_i = AR_{i-1}$, also does not depend on any semi-inverse matrices. Since the dimensions of the map of the matrix are equal to its rank, then it is thereby shown that the rank of the matrix A_i and, therefore by virtue of the equality (1.4.21), also its defect, are independent of the particular choice of semi-inverse matrices, in terms of which it is expressed by the formula (1.5.11).

From this, in view of Lemmas 1.5.2 and 1.5.3, it readily follows that, irrespective of the particular choice of semi-inverse matrices, the equality $P_r B_r = 0$ is satisfied at the same value of r. This allows the following definitions to be introduced.

Definition 1.5.2   The number k that is equal to the smallest of the non-negative integer numbers i, at which $P_i B_i = 0$, is said to be the right-hand index of the pair of matrices (A, B).

Another (different from (1.5.17)) pair of chains, which are closely associated with the chains (1.5.17) for the pair of matrices (A^*, B^*), can be compared with the pair of matrices (A, B). For this purpose, it is sufficient to construct the chains (1.5.17) for the pair of matrices (A^*, B^*) and then, by taking account of Lemma 1.1.8, to go over to conjugate matrices. As a result, we obtain the chains

$$\tilde{A}_{i+1} = \tilde{R}_i A, \quad \tilde{B}_{i+1} = \tilde{R}_i B, \quad i = 0, 1, \dots,$$

in which

$$\tilde{R} = E - \tilde{T}_i \tilde{T}_i^-, \qquad \tilde{T}_i = B\tilde{P}_i,$$
$$\tilde{P}_i = E - \tilde{A}_i^- \tilde{A}_i, \qquad \tilde{A}_0 = A, \qquad \tilde{B}_0 = B$$

(the chain $\tilde{A}_i$ can also be written as $\tilde{A}_{i+1} = \tilde{R}_i \tilde{A}_i$), or, for greater visualization, as

$$\tilde{A}_{i+1} = [E - (B - B\tilde{A}_i^- \tilde{A}_i)(B - B\tilde{A}_i^- \tilde{A}_i)^-]A,$$
$$\tilde{B}_{i+1} = [E - (B - B\tilde{A}_i^- \tilde{A}_i)(B - B\tilde{A}_i^- \tilde{A}_i)^-]B.$$

Definition 1.5.3 The left-hand index of the pair of matrices (A, B) is said to be the right-hand index of the pair of matrices (A^*, B^*), i.e. the smallest number $s \geqslant 0$ at which $\tilde{B}_s \tilde{P}_s = 0$.

Obviously, if a pair of matrices (A, B) is quite perfect, then its right-hand index is zero, and if the right-hand index of the pair of matrices (A, B) equals k, then since $(E - A_k A_k^-)B_k \equiv P_k B_k = 0$, the pair of matrices (A_k, B_k) is quite perfect.

The notions of the right- and left-hand indices of the pair of matrices (A, B) are closely related to the notions of minimum indices of the bundle $\lambda A + B$ in the columns and rows (see [1, p. 342]). However, we shall not investigate these relationships here and note only that the notions of the canonical form and of minimum indices of the bundle were used in [7, 8].

Let us return to equation (1.5.1) and let us try to answer the second question posed at the beginning of this section, without assuming that the pair of matrices (A, B) is quite perfect.

Let us first prove the following lemma.

Lemma 1.5.4 At any number λ and at any $i = 0, 1, \ldots$ the equation

$$A(\lambda x) = Bx \qquad (1.5.25)$$

is equivalent to the system

$$x = R_{i-1} x, \qquad (1.5.26)$$

$$A_i(\lambda x) = B_i x \qquad (1.5.27)$$

(when $i = 0$, one has to put $R_{-1} = E$).

Proof Let x be the solution of equation (1.5.25) and $i > 0$ (when $i = 0$, the assertion of the lemma is obvious). By multiplying the equality (1.5.25) by the matrix P (see designations (1.5.16)), by virtue of the determining property of the semi-inverse matrices, we obtain $T_0 x = 0$ and, consequently, $x = R_0 x$. Substitution of this expression for x into the left-hand side of the equality (1.5.25) yields $A_1(\lambda x) = Bx$.

If the obtained equality is multiplied by the matrix P_1, then, as above, the relationship $x = R_1 x$ arises, the substitution of which into the left-hand side of (1.5.25) leads to the equality $A_2(\lambda x) = Bx$.

By continuing this process, at the ith step we shall have the equality (1.5.26), the substitution of which into both the left- and right-hand sides of equation (1.5.25) leads to the equality (1.5.27). Thus, any solution of equation (1.5.25) satisfies the system (1.5.26), (1.5.27).

Now let x be the solution of the system (1.5.26), (1.5.27). Then, in view of (1.6.17), the equalities

$$A(\lambda R_{i-1} x) = B(R_{i-1} x), \qquad R_{i-1} x = x,$$

are valid, from which it obviously follows that x satisfies equation (1.5.25). The lemma is proved.

If k is the right-hand index of the pair of matrices (A, B), then when $i = k$ the pair of matrices in equation (1.5.27) becomes quite perfect; therefore, if λ is not an eigen-number of the matrix $A_k^- B_k$, the general solution of equation (1.5.27), according to formula (1.5.7), can be represented as

$$x = (\lambda E - A_k^- B_k)^{-1}(E - A_k^- A_k)v, \qquad (1.5.28)$$

where v is an arbitrary vector. By substituting (1.5.28) into the right-hand side of (1.5.26), for the general solution of the system (1.5.26), (1.5.27) (and therefore, by virtue of Lemma 1.5.4, also for equation (1.5.25)), we obtain the formula

$$x = R_{k-1}(\lambda E - A_k^- B_k)^{-1}(E - A_k^- A_k)v, \qquad (1.5.29)$$

where v is an arbitrary vector.

Through direct substitution, in view of the equalities $R_{k-1}^2 = R_{k-1}$ and $(E - A_k A_k^-)B_k = 0$ and formulae (1.5.17), it is easy to demonstrate that the right-hand side of (1.5.29) is, indeed, the solution of equation (1.5.25) at any vector v.

Now it is possible to formulate the lemma which gives the answer to question (2) in the case of an arbitrary pair of matrices.

Lemma 1.5.5 Let k be the right-hand index of the pair of matrices (A, B) and the number λ is not an eigen-number of the matrix $A_k^- B_k$. Then the equality

$$\det(\lambda A - B) = 0 \qquad (1.5.30)$$

is valid if and only if

$$R_{k-1}(E - A_k^- A_k) = 0. \qquad (1.5.31)$$

Proof We avail ourselves of the known equality:

$$\mathcal{M}(\lambda E - \mathcal{N}\mathcal{M})^{-1} = (\lambda E - \mathcal{M}\mathcal{N})^{-1}\mathcal{M}, \qquad (1.5.32)$$

which must be understood so that, if the left-hand side of the equality (1.5.32) is meaningful, then its right-hand side is also simultaneously meaningful and they are equal.

Putting $\mathcal{M} = R_{k-1}, \mathcal{N} = A_k^- B_k$ in (1.5.32) (taking into consideration that $R_{k-1}^2 = R_{k-1}$ and, therefore,

$$\mathcal{N}\mathcal{M} = A_k^- B_k R_{k-1} = A_k^- B R_{k-1}^2 = A_k^- B R_{k-1} = A_k^- B_k),$$

we obtain:

$$R_{k-1}(\lambda E - A_k^- B_k)^{-1} = (\lambda E - R_{k-1} A_k^- B_k)^{-1} R_{k-1}.$$

But then formula (1.5.29) can be written as

$$x = (\lambda E - R_{k-1} A_k^- B_k)^{-1} R_{k-1}(E - A_k^- A_k)v. \qquad (1.5.33)$$

From formula (1.5.33) (as a consequence of the arbitrariness of v) it follows that equation (1.5.25) has a unique (zero) solution (in other words, for the pair of matrices (A, B), the equality (1.5.30) is valid) if and only if the equality (1.5.31) is satisfied. The lemma is proved.

An application of the notion of a quite perfect pair of matrices may be found in [9, 10].

1.6. Perfect pairs of matrices

Definition 1.6.1 The pair of matrices (A, B) is said to be perfect if the equalities

$$(E - AA^-)B(A^-B)^i(E - A^-A) = 0, \quad i = 0, 1, \dots. \tag{1.6.1}$$

are satisfied.

In the next section, in connection with considering perfect pairs of variable matrices, it will be shown that Definition 1.6.1 does not presuppose any particular choice of the semi-inverse matrix to the matrix A; the validity of the equalities (1.6.1) with some single semi-inverse matrix also implies their validity with any other one.

Taking this remark into account, it is possible to obtain another (rank) characteristic of perfect pairs of matrices.

Theorem 1.6.1 The pair of matrices (A, B) is perfect if and only if at any numbers α the inequality

$$\text{rank}\, A \geqslant \text{rank}\, (A - \alpha B) \tag{1.6.2}$$

is satisfied.

Proof Before proving the theorem, we note that the system of equalities (1.6.1) is valid if and only if for all numbers α, at which the matrix $E - \alpha A^- B$ is inverse, the equality

$$(E - AA^-)B(E - \alpha A^- B)^{-1}(E - A^- A) = 0 \tag{1.6.3}$$

is satisfied. This may be demonstrated by applying the apparatus based on using matrix series and functions of matrices (see [1, pp. 101–123]).

In addition we note that the case when the rows and columns of the matrix A are linearly independent is trivial because in this case, according to Lemmas 1.4.13 and 1.4.14, the pair of matrices (A, B) is perfect and the inequality (1.6.2) is obvious. Therefore we shall prove Theorem 1.6.1 under the assumption that $\text{rank}\, A < \min\,(m, n)$.

Let us take, as the semi-inverse matrix A^- in the equality (1.6.3), the matrix (1.1.12), which is obtained from A with the help of Corollary 1.1.3. Then we can write the following block representations:

$$A = P\begin{pmatrix} A_{11} & A_{12} \\ A_{21} & A_{22} \end{pmatrix}Q, \qquad A^- = Q*\begin{pmatrix} A_{11}^{-1} & 0 \\ 0 & 0 \end{pmatrix}P*, \qquad B = P\begin{pmatrix} B_{11} & B_{12} \\ B_{21} & B_{22} \end{pmatrix}Q,$$

$$(1.6.4)$$

where $A_{22} = A_{21}A_{11}^-A_{12}$ and rank $A = $ rank A_{11}.

If $\alpha \neq 0$, then by substituting the representations (1.6.4) into the equality (1.6.3), we find that the equality (1.6.3) is equivalent to the equality

$$A_{22} - \alpha B_{22} = (A_{21} - \alpha B_{21})(A_{11} - \alpha B_{11})^{-1}(A_{12} - \alpha B_{12}),$$

which is valid according to Lemma 1.1.5 if and only if

$$\text{rank}(A - \alpha B) = \text{rank}(A_{11} - \alpha B_{11}). \tag{1.6.5}$$

The equality (1.6.5) can also be written as

$$\text{rank } A = \text{rank}(A - \alpha B) \tag{1.6.6}$$

because under the assumptions made with respect to α, the matrix $A_{11} - \alpha B_{11}$, together with the matrix A_{11}, is nonsingular and, according to the condition of the representations (1.6.4), rank $A_{11} = $ rank A.

Thus, it has been shown that, if $\alpha \neq 0$ and α^{-1} is not an eigen-number of the matrix $A^- B$ (or, equivalently, of the matrix $A_{11}^{-1}B_{11}$), then the equalities (1.6.3) and (1.6.6) are equivalent. In this case (as a consequence of the continuity of the left-hand side of the equality (1.6.3), when $\alpha \to 0$) from (1.6.6) it follows that the equality (1.6.3), when $\alpha = 0$, is also valid. All this, taken together, means that the validity of the equality (1.6.6) at all α, for which α^{-1} is not an eigen-number of the matrix $A^- B$, is a necessary and sufficient condition for the perfection of the pair of matrices (A, B).

In order to complete the proof of Theorem 1.6.1, it remains to demonstrate that, if $\alpha_0 \neq 0$ and α_0^{-1} coincides with one of the eigen-numbers of the matrix $A^- B$, then in the case when the pair of matrices (A, B) is perfect, the inequality

$$\text{rank } A \geqslant \text{rank}(A - \alpha_0 B)$$

holds.

Indeed, let us assume the opposite case:

$$\text{rank } A < \text{rank}(A - \alpha_0 B).$$

Then in the matrix $A - \alpha_0 B$ there is a zero-different minor, the dimension of which is larger than the rank of the matrix A. This minor, while being the value of the corresponding polynomial $P(\alpha)$ at the point $\alpha = \alpha_0$, is different from zero in some of its neighbourhood as well. But then throughout this neighbourhood

$$\text{rank } A < \text{rank}(A - \alpha B),$$

which contradicts the equality (1.6.6). Theorem 1.6.1 is proved.

The following theorem is of great value when investigating and solving singular systems of ordinary differential equations.

Theorem 1.6.2 For any pair of matrices (A, B), there exists a constant α at which the pair of matrices $(A - \alpha B, B)$ is perfect.

Proof Let $r(\beta) = \text{rank}(A - \beta B)$. The function $r(\beta)$ can take on only a finite number of values because the rank of the matrix does not exceed the smallest of its dimensions and is an integer number. But then the maximum of the function $r(\beta)$ is reached at a certain $\beta = \alpha$ and at any β the inequality

$$\text{rank}(A - \alpha B) \geqslant \text{rank}(A - \beta B) \qquad (1.6.7)$$

holds.

In particular, when $\beta = \alpha + \gamma$, where γ is an arbitrary number, the inequality

$$\text{rank}(A - \alpha B) \geqslant \text{rank}[(A - \alpha B) - \gamma B] \qquad (1.6.8)$$

follows from the inequality (1.6.7), which is valid at any γ. But then, according to Theorem 1.6.1, the pair of matrices $(A - \alpha B, B)$ is perfect. Theorem 1.6.2 is proved.

Remark 1.6.1 Since for a perfect pair of matrices in the inequality (1.6.8), at almost all numbers γ (with the exception of their finite set only, however), the equality

$$\text{rank}(A - \alpha B) = \text{rank}[(A - \alpha B) - \gamma B]$$

is satisfied (see (1.6.6)), then by virtue of (1.6.8)

$$\text{rank}[A - (\alpha + \gamma)B] \geqslant \text{rank}[A - (\alpha + \gamma)B - \delta B]$$

at almost all γ and all δ.

From here, in view of Theorem 1.6.1 and of the fact that $\alpha + \gamma$ together with γ runs over almost all numbers, it follows that almost all numbers α (with the exception of their finite set only, however) are suitable for the perfection of the pair of matrices $(A - \alpha B, B)$.

We show the usefulness of the notion of perfect pairs of matrices by considering an example of the solution of questions (1) and (2) formulated at the beginning of the previous section.

Let us formulate, first of all, the following lemma.

Lemma 1.6.1 If x is the solution of the equation

$$A(\lambda x) = Bx, \qquad (1.6.9)$$

then there exists a vector u such that the pair (x, u) is the solution of the system

$$\lambda x = A^- Bx + u, \qquad (1.6.10)$$

$$(E - AA^-)Bx = 0, \qquad (1.6.11)$$

$$Au = 0. \qquad (1.6.12)$$

Conversely, if the pair (x, u) is the solution of the system $(1.6.10)$–$(1.6.12)$, then the vector x is the solution of equation $(1.6.9)$, and

$$u = (E - A^- A)(\lambda x). \qquad (1.6.13)$$

Proof Let x be the solution of equation $(1.6.9)$. We form a vector $(1.6.13)$. Through a direct substitution, in view of the equality $(1.6.9)$ and the property $AA^- A = A$, it is easy to show that the pair (x, u) satisfies equations $(1.6.10)$ and $(1.6.12)$. Besides, by premultiplying the equality $(1.6.9)$ by the matrix $E - AA^-$, we find that x satisfies equation $(1.6.11)$ as well.

Conversely, if the pair (x, u) is the solution of the system $(1.6.10)$–$(1.6.12)$, then by multiplying the equality $(1.6.10)$ by the matrix A and taking in this case the equalities $(1.6.11)$ and $(1.6.12)$ into account, we find that the vector x satisfies equation $(1.6.9)$. And finally, by multiplying the equality $(1.6.10)$ by the matrix $E - A^- A$, by virtue of $(1.6.11)$ and $(1.6.12)$, we get

$$(E - A^- A)(\lambda x) = (E - A^- A)A^- Bx + u = A^-(E - AA^-)Bx + u = u,$$

i.e. the equality $(1.6.13)$ holds. The lemma is thereby proved.

If λ is not an eigen-number of the matrix AB, then Lemma 1.6.1 permits us to solve the homogeneous system $(1.6.9)$ for x and thereby to determine all eigenvectors of the bundle $B - \lambda A$ which correspond to this λ. In fact, by solving equation $(1.6.10)$ for x, we obtain:

$$x = (\lambda E - A^- B)^{-1} u. \qquad (1.6.14)$$

Next, on substituting $(1.6.14)$ into $(1.6.11)$ we have

$$(E - AA^-)B(\lambda E - A^- B)^{-1} u = 0, \qquad (1.6.15)$$

and therefore the vector u satisfies the system $(1.6.15)$, $(1.6.12)$. But then by Lemma 1.4.1

$$u = (E - A^- A)(E - Q^- Q)v, \qquad (1.6.16)$$

where

$$Q = (E - AA^-)B(\lambda E - A^- B)^{-1}(E - A^- A) \qquad (1.6.17)$$

(cf. $(1.6.3)$), and v is an arbitrary vector. Finally, substitution of $(1.6.16)$ into $(1.6.14)$ gives the required general solution of the system $(1.6.9)$:

$$x = (\lambda E - A^- B)^{-1}(E - A^- A)(E - Q^- Q)v. \qquad (1.6.18)$$

If the pair of matrices (A, B) is perfect, i.e. equalities $(1.6.1)$ are satisfied, then as a consequence of the fact that the matrix $(\lambda E - A^- B)$ is a polynomial of the matrix $A^- B$, from $(1.6.17)$ it follows that $Q = 0$, and formula $(1.6.18)$ assumes the form

$$x = (\lambda E - A^- B)^{-1}(E - A^- A)v. \tag{1.6.19}$$

In other words, in the case of a perfect pair of matrices (A, B) and with λ, which is not an eigen-number of the matrix $A^- B$, the solution of equation (1.6.9) is represented by the formula (1.6.14), where u is an arbitrary solution of the equation $Au = 0$. This simple remark turns out to be important when constructing solutions of degenerate systems of ordinary differential equations.

To summarize, let us formulate a theorem which is the consequence of the arbitrariness of the vector v in formula (1.6.19).

Theorem 1.6.3 If the pair of matrices (A, B) is perfect and λ is not an eigen-number of the matrix $A^- B$, then the equality

$$\operatorname{def}(B - \lambda A) = 0$$

is valid if and only if $E - A^- A = 0$ (in other words, if the columns of the matrix A are linearly independent (see Corollary (1.4.13))).

1.7. Perfect pairs and triplets of variable matrices

Obviously, the system of equalities (1.6.1) which define the perfection of the pair of matrices (A, B) is equivalent to the identity

$$(E - AA^-)BX(t)X^{-1}(s)(E - A^- A) = 0, \quad \alpha \leqslant t, s \leqslant \beta, \alpha \neq \beta, \tag{1.7.1}$$

in which the matrix $X(t)$ is the solution of Cauchy's problem

$$X'(t) = A^- BX(t), \quad \alpha \leqslant t \leqslant \beta,$$
$$X(\alpha) = E,$$

i.e.

$$X(t) = \exp(tA^- B),$$
$$X(t)X^{-1}(s) = \exp[(t - s)A^- B].$$

This factor induces us to suggest one of the possible generalizations of the notion of a perfect pair of matrices for the case when the matrices are variable.

Bearing in mind the applications to the study and construction of boundary-value problems for degenerate systems, we shall formulate the corresponding definition for the triplet of matrices.

Definition 1.7.1 A triplet of continuous matrices $(A(t), B(t), C(t))$ is said to be perfect on the segment $\alpha \leqslant t \leqslant \beta$ if, for the matrix $A(t)$, there exists a continuous semi-inverse matrix $A^-(t)$ for which the identities

$$[E - A(t)A^-(t)]B(t)X(t)X^{-1}(s)[E - A^-(s)A(s)] \equiv 0, \tag{1.7.2}$$

$$C(t)X(t)X^{-1}(s)[E - A^-(s)A(s)] \equiv 0, \quad s, t \in [\alpha, \beta], \tag{1.7.3}$$

are satisfied, where $X(t)$ is the solution of Cauchy's problem

$$X'(t) = A^-(t)B(t)X(t), \qquad X(\alpha) = E \tag{1.7.4}$$

(clearly, for the constant matrices A and B, the identity (1.7.2) coincides with the identity (1.7.1)).

An important role in the study of perfect triplets of matrices is played by the following two lemmas (cf. Bellman's lemma [11, p. 188]).

Lemma 1.7.1   If the elements of the matrices $\mathcal{M}(s_1, s)$ and $Z(t, s_1)$ are continuous in both variables ($\alpha \leqslant t \leqslant s_1 \leqslant \beta$, $\alpha \leqslant s_1 \leqslant s \leqslant \beta$) and the identity

$$Z(t, s) \equiv \int_t^s Z(t, s_1) \mathcal{M}(s_1, s) \, ds_1, \qquad \alpha \leqslant t \leqslant s \leqslant \beta, \tag{1.7.5}$$

is satisfied, then it is necessary that

$$Z(t, s) \equiv 0, \qquad \alpha \leqslant t \leqslant s \leqslant \beta. \tag{1.7.6}$$

Proof   From the identity (1.7.5) for any $\varepsilon > 0$, we have

$$\| Z(t, s) \| \leqslant \varepsilon + K \int_t^s \| Z(t, s_1) \| \, ds_1, \tag{1.7.7}$$

where K is a constant and $K \geqslant \| \mathcal{M}(s_1, s) \|$. Therefore, when $s \geqslant t$

$$K \| Z(t, s) \| \Big/ \left[\varepsilon + K \int_t^s \| Z(t, s_1) \| \, ds_1 \right] \leqslant K,$$

or (at a fixed t)

$$\frac{d}{ds} \ln \left[\varepsilon + K \int_t^s \| Z(t, s_1) \| \, ds_1 \right] \leqslant K.$$

From here, after integration from t to s, we obtain:

$$-\ln \varepsilon + \ln \left[\varepsilon + K \int_t^s \| Z(t, s_1) \| \, ds_1 \right] \leqslant (s - t)K$$

and, therefore,

$$\varepsilon + K \int_t^s \| Z(t, s_1) \| \, ds_1 \leqslant \varepsilon \, e^{(s-t)K}.$$

By comparing this inequality with inequality (1.7.7), we obtain:

$$\| Z(t, s) \| \leqslant \varepsilon \, e^{(s-t)K}, \qquad \beta \geqslant s \geqslant t \geqslant \alpha. \tag{1.7.8}$$

Since ε is arbitrary, (1.7.6) follows from (1.7.8). Lemma 1.7.1 is proved.

Proceeding along similar lines, we prove

Lemma 1.7.2 If the elements of the matrices $\mathcal{M}(t, s_1)$ and $Z(s_1, s)$ are continuous in both variables $(\alpha \leqslant s_1 \leqslant t \leqslant \beta, \alpha \leqslant s \leqslant s_1 \leqslant \beta)$ and the identity

$$Z(t, s) = \int_s^t \mathcal{M}(t, s_1) Z(s_1, s) \, ds_1, \quad \alpha \leqslant s \leqslant t \leqslant \beta,$$

is satisfied, then it is necessary that $Z(t, s) \equiv 0$ when $\alpha \leqslant s \leqslant t \leqslant \beta$.

We shall also need the following lemma.

Lemma 1.7.3 The general (continuous on $\alpha \leqslant t \leqslant \beta$) solution $Y(t)$ of the matrix equation $A(t) Y(t) A(t) = A(t)$ in which the matrix $A(t)$ is continuous, is written as

$$Y(t) = A^-(t) + [E - A^-(t)A(t)]U(t) + V(t)[E - A(t)A^-(t)], \quad (1.7.9)$$

where $A^-(t)$ is a certain continuous semi-inverse matrix to the matrix $A(t)$ (if, of course, it exists), and $U(t)$ and $V(t)$ are arbitrary (continuous on $[\alpha, \beta]$) matrices of appropriate dimensions. In other words, all continuous semi-inverse matrices to the continuous matrix $A(t)$ are obtained by formula (1.7.9).

Proof Clearly, the matrix (1.7.9) is continuous and satisfies the equation

$$A(t) Y(t) A(t) = A(t). \quad (1.7.10)$$

If, however, the matrix $Y(t)$ is continuous and satisfies equation (1.7.10), then it is representable as (1.7.9) when

$$U(t) = Y(t)A(t)A^-(t) - A^-(t), \quad V(t) = Y(t),$$

in which case, as a consequence of the continuity of $Y(t)$, $A(t)$, and $A^-(t)$, the matrices $U(t)$ and $V(t)$ are continuous. Lemma 1.7.3 is proved.

The next theorem shows the independence of Definition 1.7.1 and, therefore, of Definition 1.6.1 of the particular choice of the continuous semi-inverse matrix $A^-(t)$.

Theorem 1.7.1 If the system of identities (1.7.2), (1.7.3) holds in the case of a certain single continuous semi-inverse matrix $A^-(t)$ to matrix $A(t)$, then it holds also in the case of the replacement in (1.7.2)–(1.7.4) of the matrix $A^-(t)$ with any other continuous semi-inverse matrix $Y(t)$ to the matrix $A(t)$.

Proof It is easy to see that the system of identities (1.7.2), (1.7.3) can be written as

$$[E - A(t)A^-(t)]B(t)Z^1(t, s) \equiv 0, \quad (1.7.11)$$

$$C(t)Z^1(t, s) \equiv 0, \quad \alpha \leqslant t, s \leqslant \beta, \quad (1.7.12)$$

where $Z^1(t, s)$ is the solution of Cauchy's problem:

$$Z_t^1(t, s) = A^-(t)B(t)Z^1(t, s), \quad Z^1(s, s) = E - A^-(s)A(s). \quad (1.7.13)$$

In order to prove the theorem, one has to make sure that from the validity of the identities (1.7.11) and (1.7.12) the validity of the identities

$$[E - A(t)Y(t)]B(t)Z^2(t, s) \equiv 0, \tag{1.7.14}$$

$$C(t)Z^2(t, s) \equiv 0, \quad \alpha \leqslant t, s \leqslant \beta, \tag{1.7.15}$$

follows, where $Z^2(t, s)$ is the solution of the problem

$$Z_t^2(t, s) = Y(t)B(t)Z^2(t, s), \qquad Z^2(s, s) = E - Y(s)A(s), \tag{1.7.16}$$

and $Y(t)$ is a continuous matrix obtainable from formula (1.7.9).

By writing the problem (1.7.16) as

$$Z_t^2(t, s) = A^-(t)B(t)Z^2(t, s) + [Y(t) - A^-(t)]B(t)Z^2(t, s),$$
$$Z^2(s, s) = E - Y(s)A(s) \tag{1.7.17}$$

and applying it to the inhomogeneous equation (1.7.17), the known formula for the solution of Cauchy's problem, we obtain:

$$Z^2(t, s) = X(t)X^{-1}(s)[E - Y(s)A(s)]$$
$$+ \int_s^t X(t)X^{-1}(\tau)[Y(\tau) - A^-(\tau)]B(\tau)Z^2(\tau, s)\,d\tau, \tag{1.7.18}$$

where $X(t)$ is the solution of the problem (1.7.4).

Next, by substituting expression (9) instead of the matrix $Y(t)$ into (1.7.18), in view of the fact that $X(t)X^{-1}(s)[E - A^-(s)A(s)] = Z^1(t, s)$, we have

$$Z^2(t, s) = Z^1(t, s)[E - U(s)A(s)] + \int_s^t Z^1(t, \tau)U(\tau)B(\tau)Z^2(\tau, s)\,d\tau$$
$$+ \int_s^t X(t)X^{-1}(\tau)V(\tau)[E - A(\tau)A^-(\tau)]B(\tau)Z^2(\tau, s)\,d\tau. \tag{1.7.19}$$

By multiplying equality (1.7.19) by the matrix $[E - A(t)A^-(t)]B(t)$, by virtue of (1.7.11), we obtain:

$$[E - A(t)A^-(t)]B(t)Z^2(t, s) = \int_s^t [E - A(t)A^-(t)]B(t)X(t)X^{-1}(\tau)V(\tau)$$
$$\times [E - A(\tau)A^-(\tau)]B(\tau)Z^2(\tau, s)\,d\tau,$$

from which, on the basis of Lemma 1.7.2 for $\alpha \leqslant s \leqslant t \leqslant \beta$, the identity

$$[E - A(t)A^-(t)]B(t)Z^2(t, s) \equiv 0 \tag{1.7.20}$$

follows, and after multiplying by the matrix $[E - A(t)Y(t)]$, the identity

$$[E - A(t)Y(t)]B(t)Z^2(t, s) \equiv 0$$

follows as well. Thus, the identity (1.7.14) for $\alpha \leqslant s \leqslant t \leqslant \beta$ is proved.

Let us now prove the validity of the identity (1.7.14) for $s \geqslant t$. For this purpose, we note that

$$[E - A(t)A^-(t)]B(t)Z^1(t, s) = T^1(t, s)[E - A^-(s)A(s)], \qquad (1.7.21)$$

$$[E - A(t)Y(t)]B(t)Z^2(t, s) = T^2(t, s)[E - Y(s)A(s)], \qquad (1.7.22)$$

where $T^1(t, s)$ and $T^2(t, s)$ are, respectively, the solutions of the problems

$$T_s^1(t, s) = - T^1(t, s)A^-(s)B(s), \qquad T^1(t, t) = [E - A(t)A^-(t)]B(t),$$

and

$$T_s^2(t, s) = - T^2(t, s)Y(s)B(s), \qquad T^2(t, t) = [E - A(t)Y(t)]B(t).$$

For the matrix $T^2(t, s)$, the relationship

$$T^2(t, s) = [E - A(t)V(t)]T^1(t, s) - \int_t^s T^2(t, \tau)V(\tau)T^1(\tau, s)\,d\tau$$

$$- \int_t^s T^2(t, \tau)[E - A^-(\tau)A(\tau)]U(\tau)B(\tau)X(\tau)X^{-1}(s)\,d\tau \qquad (1.7.23)$$

holds, where $X(s)$ is the solution of the problem (1.7.4). By multiplying equality (1.7.23) from the right by the matrix $E - A^-(s)A(s)$, in view of (1.7.21) and (1.7.11) we obtain:

$$T^2(t, s)[E - A^-(s)A(s)]$$

$$= - \int_t^s T^2(t, \tau)[E - A^-(\tau)A(\tau)]U(t)B(\tau)X(\tau)X^{-1}(s)[E - A^-(s)A(s)]\,d\tau,$$

from which, on the basis of Lemma 1.7.1 for $\alpha \leqslant t \leqslant s \leqslant \beta$, we obtain the identity

$$T^2(t, s)[E - A^-(s)A(s)] \equiv 0.$$

and, consequently (see (1.7.22)), when $\alpha \leqslant t \leqslant s \leqslant \beta$

$$[E - A(t)Y(t)]B(t)Z^2(t, s) \equiv T^2(t, s)[E - Y(s)A(s)]$$
$$\equiv T^2(t, s)[E - A^-(s)A(s)][E - U(s)A(s)] \equiv 0.$$

Thus, the identity (1.7.14) is valid at all $s, t \in [\alpha, \beta]$.

In order to verify the validity of the identity (1.7.15), we note that, by virtue of the just proven identity (1.7.14), the second integral in the equality (1.7.19) for all s, t is zero and, hence,

$$Z^2(t, s) = Z^1(t, s)[E - U(s)A(s)] + \int_s^t Z^1(t, \tau)U(\tau)B(\tau)Z^2(\tau, s)\,d\tau. \qquad (1.7.24)$$

By multiplying from the left the equality (1.7.24) by the matrix $C(t)$, and taking the identity (1.7.12) into account, we obtain the identity (1.7.15). Theorem 1.7.1 is proved.

If in the group of three matrices $(A(t), B(t), C(t))$ the matrix $C(t)$ is of zero order, then it is natural to speak not of the group of three matrices $(A(t), B(t), 0)$ but of a pair of matrices $(A(t), B(t))$. In particular, the pair of variables of the matrices $(A(t), B(t))$, for which the identity (1.7.2) is satisfied, will henceforth be called (in accordance with Definition 1.6.1) the *perfect pair of matrices*.

The following lemma allows us to formulate the definition of the perfectness of the group of three matrices in different, equivalent forms.

Lemma 1.7.4   The identities given below are equivalent:

(1) $G_1(t)G_2(s) \equiv 0$,
(2) $\Gamma_1 G_2(s) \equiv 0$,
(3) $G_1(t)\Gamma_2 \equiv 0$,
(4) $\Gamma_1\Gamma_2 = 0$.

Here $\alpha \leqslant t \leqslant \beta$, $\alpha \leqslant s \leqslant \beta$,

$$\Gamma_1 = \int_\alpha^\beta G_1^*(t)G_1(t)\,dt, \qquad \Gamma_2 = \int_\alpha^\beta G_2(s)G_2^*(s)\,ds, \qquad (1.7.25)$$

and the matrices $G_1(t)$ and $G_2(t)$ are continuous.

Proof   By applying Lemma 1.3.3 to identity (1) (with fixed s), we obtain the equivalence of (1) and (2). Similarly, by applying Lemma 1.3.3 to conjugate identity (1) (with fixed t), we obtain the equivalence of (1) and (3), etc.

If, for example,

$$G_1(t) = [E - A(t)A^-(t)]B(t)X(t), \qquad G_2(s) = X^{-1}(s)[E - A^-(s)A(s)], \qquad (1.7.26)$$

then the validity of identity (1) means the perfectness of the pair of matrices $(A(t), B(t))$, and Lemma 1.7.4 implies that the validity of each of the identities (2)–(4) means also the perfectness of the pair of matrices $(A(t), B(t))$. In this case, identities (1.7.2) and (1.7.3) from the definition of the perfect group of three matrices can be written, for example, accordingly as

$$\Gamma_1\Gamma_2 = 0, \qquad C(t)X(t)\Gamma_2 = 0, \qquad (1.7.27)$$

where Γ_1 and Γ_2 are obtainable from formulae (1.7.25) and (1.7.26).

By solving the second equation from (1.7.27) for the matrix $C(t)$, we obtain:

$$C(t) = Z(t)(E - \Gamma_2\Gamma_2^-)X^{-1}(t), \qquad (1.7.28)$$

from which follows the theorem useful for formulating boundary-value problems for singular systems.

Theorem 1.7.2   The matrix $C(t)$ can enter the composition of the perfect group of three matrices $(A(t), B(t), C(t))$ if and only if it is represented as in (1.7.28), where $Z(t)$ is a certain matrix, and the matrix Γ_2 is defined by formulae (1.7.25) and (1.7.26), in which $X(t)$ is the solution of the problem (1.7.4).

Corollary 1.7.1 The group of three matrices $(A(t), B(t), E)$ is perfect if and only if the columns of the matrix $A(t)$ are linearly independent.

Proof The proof of this corollary can be performed using formula (1.7.28) as well as directly, by putting $s = t$ in the identities (1.7.2) and (1.7.3) and by applying Lemma 1.4.13.

1.8. The inverse Drasin matrix

The inverse Drasin matrix [12] is defined only for square matrices.

Definition 1.8.1 The matrix $A^{\mathscr{D}}$ that satisfies the system of equations

(1) $AA^{\mathscr{D}} = A^{\mathscr{D}}A$,
(2) $A^{\mathscr{D}}AA^{\mathscr{D}} = A^{\mathscr{D}}$, $\qquad\qquad$ (1.8.1)
(3) $(E - A^{\mathscr{D}}A)A^k = 0$,

is said to be the inverse Drasin matrix to the matrix A, where k is the index of the matrix A, i.e. the smallest of the non-negative integer numbers for which rank $A^{k+1} = $ rank A^k.

The equality $k = 0$ means that the matrix A is nonsingular. In this case the system (1.8.1) is satisfied by the matrix $A^{\mathscr{D}} = A^{-1}$. When $k = 1$, the matrix $A^{\mathscr{D}}$, by virtue of (1)–(3), coincides with one of the inverse semi-reciprocal matrices $A^\sim$ (see (1.1.14). If, however, the matrix A is self-conjugate, then the comparison of (1.8.1) with (1.2.19) leads to the equality $A^{\mathscr{D}} = A^+$.

The simple meaning of the inverse Drasin matrix is disclosed by the following theorem.

Theorem 1.8.1 The inverse Drasin matrix $A^{\mathscr{D}}$ to an arbitrary $(n \times n)$ matrix A exists, is unique, and allows for a representation which is constructed as follows. Let

$$A = N\begin{pmatrix} J_0 & 0 \\ 0 & J_1 \end{pmatrix}N^{-1} \qquad (1.8.2)$$

be a Jordan representation of the matrix A, where J_0 consists of nilpotent blocks, and J_1 consists of nonsingular blocks. Then

$$A^{\mathscr{D}} = N\begin{pmatrix} 0 & 0 \\ 0 & J_1^{-1} \end{pmatrix}N^{-1}. \qquad (1.8.3)$$

Proof To begin with, we note that the index k of the matrix A equals the order l of the greatest nilpotent block in its Jordan representation (1.8.2) because rank $A^{i+1} < $ rank A^i when $i < l$ and rank $A^{i+1} = $ rank A^i. Next, taking into consideration that $J_0^k = 0$, we perform the substitution of the representations

(1.8.2) and (1.8.3) into equalities (1.8.1). As a result, we make sure that equalities (1.8.1) are satisfied.

In this way one can prove the existence of the inverse Drasin matrix $A^{\mathscr{D}}$ having the representation (1.8.3).

Let us now prove that for a given matrix A two different inverse Drasin matrices $A_1^{\mathscr{D}}$ and $A_2^{\mathscr{D}}$ cannot exist. Let $R = A_1^{\mathscr{D}} - A_2^{\mathscr{D}}$. The first and third equalities from (1.8.1) yield

$$RA = AR, \qquad RA^{k+1} = 0, \tag{1.8.4}$$

and the second gives

$$(AA_1^{\mathscr{D}})^{k+1} = AA_1^{\mathscr{D}}AA_1^{\mathscr{D}}(AA_1^{\mathscr{D}})^{k-1} = AA_1^{\mathscr{D}}(AA_1^{\mathscr{D}})^{k-1} = (AA_1^{\mathscr{D}})^k = \cdots = AA_1^{\mathscr{D}}, \tag{1.8.5}$$

and similarly

$$(A_2^{\mathscr{D}}A)^{k+1} = A_2^{\mathscr{D}}A. \tag{1.8.6}$$

From (1.8.4)–(1.8.6), in view of the first and second equalities (1.8.1), it follows that

$$A_1^{\mathscr{D}} = A_1^{\mathscr{D}}AA_1^{\mathscr{D}} = (A_1^{\mathscr{D}} - A_2^{\mathscr{D}})AA_1^{\mathscr{D}} + A_2^{\mathscr{D}}A(A_1^{\mathscr{D}} - A_2^{\mathscr{D}}) + A_2^{\mathscr{D}}$$
$$= RAA_1^{\mathscr{D}} + A_2^{\mathscr{D}}AR + A_2^{\mathscr{D}} = RA^{k+1}(A_1^{\mathscr{D}})^{k+1} + (A_2^{\mathscr{D}})^{k+1}A^{k+1}R + A_2^{\mathscr{D}} = A_2^{\mathscr{D}}.$$

Theorem 1.8.1 is thereby proved.

The representation (1.8.3) can also be obtained with the help of an elementary theorem.

Theorem 1.8.2   If the matrix N is nonsingular and

$$A = NBN^{-1}, \tag{1.8.7}$$

then

$$A^{\mathscr{D}} = NB^{\mathscr{D}}N^{-1}. \tag{1.8.8}$$

The proof can be performed by substituting (1.8.7) and (1.8.8) into the equalities (1.8.1), in view of the fact that, as a consequence of the nonsingularity of the N matrix, the ranks and, hence, also the indices of the matrices A and B coincide.

As for the matrices $A(t)$, the elements of which are certain functions of the parameter $t \in [\alpha, \beta]$, the following theorem is valid.

Theorem 1.8.3   If $A(t) \in C^1[\alpha, \beta]$ and $A^{\mathscr{D}}(t) \in C^1[\alpha, \beta]$, then

$$(A^{\mathscr{D}})' = -A^{\mathscr{D}}A'A^{\mathscr{D}} + (E - A^{\mathscr{D}}A)(A^k)'(A^{\mathscr{D}})^{k+1} + (A^{\mathscr{D}})^{k+1}(A^k)'(E - A^{\mathscr{D}}A),$$

where k is the index of the matrix A.

Proof First of all, taking the equalities (1.8.1) into consideration, we obtain:

$$(A^{\mathscr{D}})' = (A^{\mathscr{D}}AA^{\mathscr{D}})' = (A^{\mathscr{D}}A)'A^{\mathscr{D}} + A^{\mathscr{D}}A(A^{\mathscr{D}})' + A^{\mathscr{D}}A'A^{\mathscr{D}} - A^{\mathscr{D}}A'A^{\mathscr{D}}$$
$$= (A^{\mathscr{D}}A)'A^{\mathscr{D}} + A^{\mathscr{D}}(AA^{\mathscr{D}})' - A^{\mathscr{D}}A'A^{\mathscr{D}},$$

i.e.

$$(A^{\mathscr{D}})' = -A^{\mathscr{D}}A'A^{\mathscr{D}} + (A^{\mathscr{D}}A)'A^{\mathscr{D}} + A^{\mathscr{D}}(A^{\mathscr{D}}A)'. \qquad (1.8.9)$$

Furthermore, since $(E - A^{\mathscr{D}}A)A^k = 0$, we obtain:

$$0 = [(E - A^{\mathscr{D}}A)A^k]' = -(A^{\mathscr{D}}A)'A^k + (E - A^{\mathscr{D}}A)(A^k)',$$

i.e.

$$(A^{\mathscr{D}}A)'A = (E - A^{\mathscr{D}}A)(A^k)'. \qquad (1.8.10)$$

If the equality (1.8.10) is now multiplied from the right by the matrix $(A^{\mathscr{D}})^{k+1}$, then since $A^k(A^{\mathscr{D}})^{k+1} = A^{\mathscr{D}}$, we obtain:

$$(A^{\mathscr{D}}A)'A^{\mathscr{D}} = (E - A^{\mathscr{D}}A)(A^k)'(A^{\mathscr{D}})^{k+1}. \qquad (1.8.11)$$

In a similar way, one can obtain:

$$A^{\mathscr{D}}(A^{\mathscr{D}}A)' = (A^{\mathscr{D}})^{k+1}(A^k)'(E - A^{\mathscr{D}}A). \qquad (1.8.12)$$

Finally, on substituting (1.8.11) and (1.8.12) into (1.8.9), we arrive at the formula (1.8.8). Theorem 1.8.3 is thereby proved.

Corollary 1.8.1 The formula

$$(A^{\mathscr{D}}A)' = (E - A^{\mathscr{D}}A)(A^k)'(A^{\mathscr{D}})^k + (A^{\mathscr{D}})^k(A^k)'(E - A^{\mathscr{D}}A) \qquad (1.8.13)$$

holds true.

Proof The proof can be performed directly, without recourse to Theorem 1.8.3. Indeed, since $(A^{\mathscr{D}}A)^2 = A^{\mathscr{D}}A$, the equality

$$(A^{\mathscr{D}}A)' = (A^{\mathscr{D}}A)'A^{\mathscr{D}}A + A^{\mathscr{D}}A(A^{\mathscr{D}}A)' \qquad (1.8.14)$$

holds. But then one can easily arrive at the formula (1.8.13) by taking into account formulae (1.8.11) and (1.8.12) and the equalities $A(A^{\mathscr{D}})^{k+1} = (A^{\mathscr{D}})^{k+1}A = (A^{\mathscr{D}})^k$.

For the subsequent treatment, we note that from formula (1.8.14) (after it has been multiplied by the matrix $A^{\mathscr{D}}A$) the useful equality

$$A^{\mathscr{D}}A(A^{\mathscr{D}}A)'A^{\mathscr{D}}A = 0 \qquad (1.8.15)$$

follows (incidentally, the equality (1.8.15) can be obtained from formula (1.8.13) as well).

The following representation of the resolvent of the matrix is also useful:

$$R(z, A) \equiv (zE - A)^{-1} = \sum_{i=1}^{k} z^{-i}A^{i-1}(E - A^{\mathscr{D}}A) - (E - zA^{\mathscr{D}})^{-1}A^{\mathscr{D}}. \qquad (1.8.16)$$

One can make sure that this representation is valid by taking into consideration equations (1.8.1) and by verifying the equality $(zE - A)(zE - A)^{-1} = E$.

If in the representation (1.8.16) $|z| \, \| A^{\mathscr{D}} \| < 1$, then the summand in (1.8.16) can be expanded into a series. As a result, in the vicinity of zero, one obtains the expansion of the resolvent into a Loran series:

$$R(z, A) = \sum_{s=1}^{k} z^{-s} A^{s-1} (E - A^{\mathscr{D}} A) - \sum_{s=0}^{\infty} z^{s} (A^{\mathscr{D}})^{s+1}. \qquad (1.8.17)$$

Note that using the known equalities

$$\frac{1}{2\pi i} \int_{K} z^{-s} \, dz = \begin{cases} 1, & s = 1, \\ 0, & s \neq 1 \end{cases} \qquad (1.8.18)$$

(the contour K is a circle with its centre at zero, passing in the negative direction), with the help of the expansion (1.8.17) it is easy to arrive at an integral representation of the inverse Drasin matrix:

$$A^{\mathscr{D}} = \frac{1}{2\pi i} \int_{K} z^{-1} R(z, A) \, dz \qquad (1.8.19)$$

(the radius of the circle K in the equality (1.8.19) must satisfy the inequality $r < \| A^{\mathscr{D}} \|^{-1}$).

Not only does formula (1.8.19) open the way to different generalizations, but it also makes it possible to establish some useful identities. For example,

$$(AB)^{\mathscr{D}} A = A(BA)^{\mathscr{D}}, \qquad (1.8.20)$$

which is immediately obtainable after integration of the obvious identity

$$\frac{1}{2\pi i} z^{-1} (zE - AB)^{-1} A = \frac{1}{2\pi i} z^{-1} A (zE - BA)^{-1}$$

(along the contour K of a sufficiently small radius), in view of formula (1.8.19).

Using inverse Drasin matrices it is also possible to give another representation of the resolvent $R(z, A)$, valid for any z. Let us do it in the following way.

Let $\lambda_1, \ldots, \lambda_m$ be different from all other eigen-numbers of the matrix A (if among the eigen-numbers $\lambda_1, \ldots, \lambda_m$ there is a zero one, then let $\lambda_1 = 0$).

Let us introduce into our treatment the matrices $A_j = A - \lambda_j E$ $(j = 1, \ldots, m)$, the indices of which will be denoted by v_j and let us prove the equality

$$\sum_{j=1}^{m} (E - A_j^{\mathscr{D}} A_j) = E. \qquad (1.8.21)$$

For this purpose, in the left-hand part of the equality (1.8.21) we substitute the matrix A for the matrix $A = NJN^{-i}$, where J is the Jordan form of the matrix A. As a result of such a substitution the proof of the equality (1.8.21) will be reduced to proving that

$$\sum_{j=1}^{m} (E - J_j^{\mathscr{D}} J_j) = E, \tag{1.8.22}$$

where $J_j = J - \lambda_j E$.

By applying Theorem 1.8.1 to the matrices J_j in (1.8.22) we will obtain that the matrices $E - J_j^{\mathscr{D}} J_j$ are diagonal matrices, and in all places, where in the Jordan form J the eigen-number λ_j of the matrix A is, unities stand in the matrix $E - J_j^{\mathscr{D}} J_j$, and all its other elements are zero. From here obviously follows the validity of the equality (1.8.22) (and, consequently, the validity of the equality (1.8.21)).

Note now that

$$R(z, A) = (zE - A)^{-1} = [(z - \lambda_j)E - A_j]^{-1} \tag{1.8.23}$$

and simultaneously as a consequence of (1.8.21) that

$$R(z, A) = \sum_{j=1}^{m} R(z, A)(E - A_j^{\mathscr{D}} A_j). \tag{1.8.24}$$

But then, upon substituting (1.8.23) into (1.8.24) we obtain:

$$R(z, A) = \sum_{j=1}^{m} R(z - \lambda_j, A_j) \cdot (E - A_j^{\mathscr{D}} A_j). \tag{1.8.25}$$

Finally, the replacement of the resolvent $R(z - \lambda_j, A_j)$ in the equality (1.8.25) with its representation (1.8.16) leads (in view of the fact that $(E - A_j^{\mathscr{D}} A_j)A_j^{\mathscr{D}} = 0$) to the formula

$$R(z, A) = \sum_{j=1}^{m} \sum_{s=1}^{v_j} \frac{1}{(z - \lambda_j)^s} A_j^{s-1}(E - A_j^{\mathscr{D}} A_j). \tag{1.8.26}$$

Formula (1.8.26) makes it possible to write the expression for the function of the matrix A. In fact, if $f(z)$ is an analytic function in a closed circle K centred on the point zero, containing all eigen-numbers of the matrix A, then, as is known (see, for example, [1, p. 119]),

$$f(A) = \frac{1}{2\pi i} \int_K f(z)R(z, A) \, dz. \tag{1.8.27}$$

But then, on substituting (1.8.26) into (1.8.27), in view of the equalities

$$\frac{1}{2\pi i} \int_K \frac{f(z)}{(z - \lambda_j)^s} \, dz = \frac{1}{(s-1)!} f^{(s-1)}(\lambda_j),$$

we obtain:

$$f(A) = \sum_{j=1}^{m} \sum_{s=1}^{v_j} \frac{1}{(s-1)!} f^{(s-1)}(\lambda_j) A_j^{s-1}(E - A_j^{\mathscr{D}} A_j). \tag{1.8.28}$$

Without going into details, we note that formula (1.8.28) also remains valid in the case if at the point $z = 0$ the analyticity of the function $f(z)$ is violated, but among the eigen-numbers of the matrix A there are no zero numbers. In particular, if the matrix A is nonsingular, then the matrix A^{-1} is also represented as (1.8.28). If, however, the matrix A is singular ($\lambda_1 = 0$) and $f(z) = z^{-1}$, then by formula (1.8.28) we obtain the inverse Drasin matrix $A^{\mathscr{D}}$. In this case it must be taken into account that only the summation over j in (1.8.28) must then be performed, starting from $j = 2$.

It is known (see, for example, [1, p. 103]) that the function $f(A)$ is a polynomial of the matrix A. An important thing for us will be the fact that the inverse Drasin matrix also has this property, namely the following theorem is valid.

Theorem 1.8.4 The inverse Drasin matrix $A^{\mathscr{D}}$ is a certain polynomial of the matrix A.

Proof Using the representation (1.8.3) and the equality $J_0^k = 0$ one can write

$$A^{\mathscr{D}} = N \begin{pmatrix} J_0^k & 0 \\ 0 & J_1^k \end{pmatrix} \begin{pmatrix} 0 & 0 \\ 0 & J_1^{-k-1} \end{pmatrix} N^{-1}, \tag{1.8.29}$$

where the matrix J_1^{-k-1}, which is a function of J_1, is a polynomial

$$J_1^{-k-1} = \sum_{i=0}^{m} \beta_i J_1^i \tag{1.8.30}$$

(m is the degree of the minimum polynomial of the matrix J_1).

On substituting (1.8.30) into (1.8.29), in view of the fact that one can substitute anything one likes into the left-hand upper corner of the third matrix on the right-hand part of the equality (1.8.29) without changing the result of the substitution (because $J_0^k = 0$), we arrive at the equality

$$A^{\mathscr{D}} = \sum_{i=0}^{m} \beta_i N \begin{pmatrix} J_0^k & 0 \\ 0 & J_1^k \end{pmatrix} \begin{pmatrix} J_0^i & 0 \\ 0 & J_1^i \end{pmatrix} N^{-1} = \sum_{i=1}^{m} \beta_i A^{k+i}.$$

Theorem 1.8.4 is thereby proved.

The following theorem gives the relation of the matrix $A^{\mathscr{D}}$ to semi-inverse matrices.

Theorem 1.8.5 [6] Let k be the index of the matrix A and $r \geqslant k$. Then

$$A^{\mathscr{D}} = A^r (A^{2r+1})^- A^r. \tag{1.8.31}$$

Proof On the basis of the representations (1.8.2) and (1.8.3) and Lemma 1.1.2 the proof of Theorem 1.8.5 is reduced to proving the equality

$$A^{\mathscr{D}} = N \begin{pmatrix} 0 & 0 \\ 0 & J_1^r \end{pmatrix} \begin{pmatrix} 0 & 0 \\ 0 & J_1^{2r+1} \end{pmatrix}^- \begin{pmatrix} 0 & 0 \\ 0 & J_1^r \end{pmatrix} N^{-1}. \tag{1.8.32}$$

In order to prove it, we note that using Lemma 1.1.1 it is easy to arrive at the equality

$$\begin{pmatrix} 0 & 0 \\ 0 & J_1^{2r+1} \end{pmatrix}^{-} = \begin{pmatrix} u_1 & u_2 \\ u_3 & J_1^{-2r-1} \end{pmatrix}, \tag{1.8.33}$$

where u_1, u_2, and u_3 are arbitrary matrices. But then, on substituting the right-hand side of the equality (1.8.33) into the right-hand side of the equality (1.8.32), we obtain (1.8.3). This completes the proof of Theorem 1.8.5.

Finally, we give another theorem which can form the basis for constructing approximate methods of calculating the matrix $A^{\mathscr{D}}$.

Theorem 1.8.6 [13] The equality

$$A^{\mathscr{D}} = \lim_{\delta \to 0} (\delta^2 E + A^{k+1})^{-1} A^k \tag{1.8.34}$$

holds true, where k is the index of the matrix A, and the estimation

$$\| A^{\mathscr{D}} - (\delta^2 E + A^{k+1})^{-1} A^k \| \leq \delta^2 \| A^{\mathscr{D}} \|^{k+2} (1 - \delta^2 \| A^{\mathscr{D}} \|^{k+1})^{-1} \tag{1.8.35}$$

holds (here $\|\cdot\|$ is the spectral norm of the matrix, and δ is real).

Proof Since the index of the matrix A^{k+1} is equal to unity, with the help of formula (1.8.16), the equalities (1.8.1), and $(A^{\mathscr{D}})^{k+1} = (A^{k+1})^{\mathscr{D}}$ one can write

$$(\delta^2 E + A^{k+1})^{-1} A^k = [E + \delta^2 (A^{\mathscr{D}})^{k+1}]^{-1} A^{\mathscr{D}}.$$

Therefore

$$A^{\mathscr{D}} - (\delta^2 E + A^{k+1})^{-1} A^k = \delta^2 [E + \delta^2 (A^{\mathscr{D}})^{k+1}]^{-1} (A^{\mathscr{D}})^{k+2},$$

from which, with sufficiently small δ, the estimation (1.8.35) and, consequently, the equality (1.8.34) follow.

1.9. The representation of the inverse Drasin matrix with the help of skeleton expansions

Let $A = B_0 \Gamma_0$ be a skeleton expansion of the $(n \times n)$ matrix A.

Let us define the matrix $A_1 = \Gamma_0 B_0$ and let us construct for it a skeleton expansion $A_1 = B_1 \Gamma_1$. We shall operate on the matrix A_1 in the same way as was done with the matrix A, namely we define the matrix $A_2 = \Gamma_1 B_1$ and construct for it the skeleton expansion $A_2 = B_2 \Gamma_2$. We shall operate on the matrix A_2 in the same way as was done with the matrix A_1, etc. As a result, we obtain a chain of matrices:

$$A_0 = A = B_0 \Gamma_0, \qquad A_i = \Gamma_{i-1} B_{i-1} = B_i \Gamma_i, \quad i = 1, 2, \ldots, \tag{1.9.1}$$

where $B_i \Gamma_i$ is a skeleton expansion of the matrix A_i.

It is easy to prove the equalities

$$AB_0 \cdots B_{i-1} = B_0 \cdots B_{i-1} A_i, \tag{1.9.2}$$

$$\Gamma_{i-1} \cdots \Gamma_0 A = A_i \Gamma_{i-1} \cdots \Gamma_0, \tag{1.9.3}$$

$$B_0 \cdots B_{i-1} \Gamma_{i-1} \cdots \Gamma_0 = A^i, \tag{1.9.4}$$

$$\Gamma_{i-1} \cdots \Gamma_0 B_0 \cdots B_{i-1} = A_i^i. \tag{1.9.5}$$

Let us prove, for example, the equality (1.9.4). In view of (1.9.1), we have

$$
\begin{aligned}
B_0 \cdots B_{i-1} \Gamma_{i-1} \cdots \Gamma_0 &= B_0 \cdots B_{i-2} A_{i-1} \Gamma_{i-2} \cdots \Gamma_0 \\
&= B_0 \cdots B_{i-2} \Gamma_{i-2} B_{i-2} \Gamma_{i-2} \cdots \Gamma_0 \\
&= B_0 \cdots B_{i-3} A_{i-2}^2 \Gamma_{i-3} \cdots \Gamma_0 = \cdots = B_0 A_1^{i-1} \Gamma_0 \\
&= B_0 (\Gamma_0 B_0)^{i-1} \Gamma_0 = (B_0 \Gamma_0)^i = A^i, \tag{1.9.6}
\end{aligned}
$$

and the equality (1.9.4) is thereby proved.

If in the chain (1.9.6) i is increased by unity, then we obtain:

$$B_0 \cdots B_{i-1} A_i \Gamma_{i-1} \cdots \Gamma_0 = A^{i+1} \tag{1.9.7}$$

(see the second and last links in (1.9.6)).

We multiply the equality (1.9.7) from the left by B_0^* and from the right by Γ_0^*. Then, as a consequence of the nonsingularity of the matrices $B_0^* B_0$ and $\Gamma_0 \Gamma_0^*$ (see the beginning of Section 1.2), we obtain:

$$B_1 \cdots B_{i-1} A_i \Gamma_{i-1} \cdots \Gamma_1 = (B_0^* B_0)^{-1} B_0^* A^{i+1} \Gamma_0^* (\Gamma_0 \Gamma_0^*)^{-1},$$

which, by virtue of Corollaries 1.2.1 and 1.2.2, yields

$$B_1 \cdots B_{i-1} A_i \Gamma_{i-1} \cdots \Gamma_1 = B_0^+ A^{i+1} \Gamma_0^+. \tag{1.9.8}$$

We shall make the same manipulation with the equality (1.9.8) (we multiply it from the left by B_1^* and from the right by Γ_1^*) and continue this process. As a result, we obtain:

$$A_i = C_i A^{i+1} D_i, \tag{1.9.9}$$

where $C_i = B_{i-1}^+ \cdots B_0^+$ and $D_i = \Gamma_0^+ \cdots \Gamma_{i-1}^+$.

Next, we make use of the known theorem about ranks: the rank of the product of matrices does not exceed the rank of any one of its co-factors. Then, from (1.9.9) we obtain:

$$\operatorname{rank} A_i \leqslant \operatorname{rank} A^{i+1},$$

and from (1.9.7):

$$\operatorname{rank} A^{i+1} \leqslant \operatorname{rank} A_i$$

and, consequently,

$$\operatorname{rank} A^{i+1} = \operatorname{rank} A_i. \tag{1.9.10}$$

Furthermore, using the same theorem and the equality (1.9.10), we arrive at the inequalities

$$\operatorname{rank} A_i \geqslant \operatorname{rank} A_{i+1}, \quad i = 0, 1, \ldots. \tag{1.9.11}$$

In fact,

$$\operatorname{rank} A_{i+1} = \operatorname{rank} A^{i+2} \leqslant \operatorname{rank} A^{i+1} = \operatorname{rank} A_i. \tag{1.9.11}$$

We denote now the order and rank of the matrix A_i by the symbols P_i and r_i, respectively. Then, according to the inequalities (1.9.11)

$$r_i \geqslant r_{i+1}, \quad i = 0, 1, \ldots. \tag{1.9.12}$$

Moreover, since in the skeleton expansion $A_{i-1} = B_{i-1}\Gamma_{i-1}$ the number of columns of the matrix B_{i-1} (as well as the number of rows of the matrix Γ_{i-1}) coincides with the rank of the matrix A_{i-1} and $A_i = \Gamma_{i-1}B_{i-1}$ (see (1.9.1)), we have

$$P_i = r_{i-1} \tag{1.9.13}$$

(when $i = 0$ we put $r_{-1} = n$, where n is the order of the matrix $A_0 = A$; then, when $i = 0$ the equality (1.9.13) will correspond to the equality $P_0 = n$).

Theorem 1.9.1 Let k be the smallest number of a set of natural numbers, at which either $r_k = 0$ or $r_{k-1} = r_k$ (such a k, by virtue of the inequalities (1.9.12), is obviously present). Then, if $r_k = 0$,

$$A^{\mathscr{D}} = 0, \tag{1.9.14}$$

and the index of the matrix A is equal to $k + 1$. If, however, $r_{k-1} = r_k$, then the matrix A_h is nonsingular and

$$A^{\mathscr{D}} = B_0 \cdots B_{k-1}(A_k^{-1})^{k+1}\Gamma_{k-1} \cdots \Gamma_0, \tag{1.9.15}$$

and the index of the matrix A equals k.

Proof Let the first possibility be realized, i.e. $r_k = 0$. Then $A_h = 0$ (the matrix rank is zero if and only if the matrix is zero) and (see (1.9.10)) $A^{k+1} = 0$. Moreover, from the assumption about the number k (see the beginning of the formulation of the theorem) it follows that

$$r_{i-1} > r_i, \quad i = 0, \ldots, k,$$

and, therefore, according to (1.9.10)

$$\operatorname{rank} A^{i+1} < \operatorname{rank} A^i, \quad i = 0, 1, \ldots, k$$

(as usual we assume $A^0 = E$). At the same time

$$\operatorname{rank} A^{k+1} = \operatorname{rank} A^{k+2},$$

because $A^{k+1} = 0$. But then, by definition of the index of the matrix (see Definition 1.8.1), the index of the matrix A equals $k + 1$. Moreover, taking into account the equality $A^{k+1} = 0$ it is easy to note that the system (1.8.1) is satisfied, for example, by the zero matrix ($A^{\mathscr{D}} = 0$). Therefore, as a consequence of the uniqueness of the solution of the system (1.8.1), $A^{\mathscr{D}} = 0$ (see (1.9.14)).

If, however, the second possibility is realized, i.e. $r_{k-1}r_k$, then by virtue of the equality (1.9.10), rank $A^{k+1} = $ rank A^k, and, at the same time, according to the assumption about the number k,

$$\text{rank } A^{i+1} < \text{rank } A^i, \quad i = 0, \dots, k-1,$$

which means that the index of the matrix A equals k (the case $k = 0$ is not excluded) then $r_0 = r_{-1} = n$ and the matrix A is nonsingular (its index, by definition, is zero).

Furthermore, since, when the second possibility is realized, $r_{k-1} = r_k$, then as a consequence of (1.9.13) we have

$$P_k = r_{k-1} = r_k,$$

i.e. the rank of the matrix A_h equals its order. But this means that the matrix A_h is nonsingular and, consequently, the matrix (1.9.15) can be constructed. On substituting the matrix (1.9.15) into the system (1.8.1) (in this case taking the equalities (1.9.2)–(1.9.5) into account), we find that the matrix (1.9.15) satisfies the system (1.8.1) and, hence, is the inverse Drasin matrix to the matrix A. The theorem is thereby proved.

1.10.   The resolving pair of matrices

Using the inverse Drasin matrix, sometimes it becomes possible to write in compact form the solution of an algebraic system of the form

$$A(\lambda x) = x + f. \tag{1.10.1}$$

Indeed, by virtue of equation (1.10.1), we have

$$x = \lambda Ax - f = \lambda A(\lambda Ax - f) - f = \lambda A(\lambda A(\lambda Ax - f) - f) - f) = \cdots.$$

On stopping at the kth link of this chain, we obtain the equality

$$x = \lambda^k A^k x - (\lambda^{k-1} A^{k-1} + \cdots + E)f.$$

By multiplying it by the matrix $(E - A^{\mathscr{D}} A)$, under the assumption that k is the index of the matrix A, we arrive at the obvious expression for the vector $(E - A^{\mathscr{D}} A)x$:

$$(E - A^{\mathscr{D}} A)x = -(E - A^{\mathscr{D}} A) \sum_{i=0}^{k-1} \lambda^i A^i f. \tag{1.10.2}$$

Furthermore, we multiply equation (1.10.1) by the matrix $A^{\mathscr{D}}$. The result of this multiplication, as a consequence of the equalities (1.8.1), can be written as

$$\lambda(A^{\mathscr{D}}Ax) = A^{\mathscr{D}}(A^{\mathscr{D}}Ax) + A^{\mathscr{D}}f.$$

Hence it follows that if λ is not the eigen-number of the matrix $A^{\mathscr{D}}$, then

$$A^{\mathscr{D}}Ax = (\lambda E - A^{\mathscr{D}})^{-1}A^{\mathscr{D}}f. \qquad (1.10.3)$$

By adding the equalities (1.10.2) and (1.10.3), we arrive at the obvious representation of the solution of the system (1.10.1):

$$x = (\lambda E - A^{\mathscr{D}})^{-1}A^{\mathscr{D}}f - (E - A^{\mathscr{D}}A)\sum_{i=0}^{k-1}\lambda^{i}A^{i}f.$$

In order to extend this method of solution to the case of the general system

$$A(\lambda x) = Bx + f, \qquad (1.10.4)$$

we introduce the notion of a resolving pair of matrices, corresponding to a given pair of matrices (A, B).

Definition 1.10.1   The pair of matrices (A^B, Y), in which

$$A^B = (YA)^{\mathscr{D}}Y \qquad (1.10.5)$$

and Y is the solution of the system

$$(E - BY)(AY)^{\mathscr{D}} = 0, \qquad (YA)^{\mathscr{D}}(E - YB) = 0, \qquad (1.10.6)$$

$$BYB = B, \qquad (1.10.7)$$

is said to be the resolving pair, cooresponding to the pair of matrixes (A, B).

Remark 1.10.1   If the solution of the system (1.10.6), (1.10.7) is determined and if the indices of the matrices AY and YA equal s and k, respectively, then the equalities (1.10.6) can be written as

$$(E - BY)(AY)^{s} = 0, \qquad (YA)^{k}(E - YB) = 0. \qquad (1.10.8)$$

Indeed, by multiplying the first equality of (1.10.6) by the matrix $(AY)^{s+1}$ from the right, and the second by the matrix $(YA)^{k+1}$ from the left, on the basis of the third property of the inverse Drasin matrix (see Definition 1.8.1), we obtain the equalities (1.10.8). Moreover, note that if the equalities (1.10.8) are satisfied, then the equalities (1.10.6) are also satisfied. This is easy to prove with the help of Theorem 1.8.5, according to which

$$(AY)^{\mathscr{D}} = (AY)^{s}[(AY)^{2s+1}]^{-}(AY)^{s},$$
$$(YA)^{\mathscr{D}} = (YA)^{k}[(YA)^{2k+1}]^{-}(YA)^{k}.$$

Upon performing the pertinent multiplications of the equalities (1.10.8), we obtain the equalities (1.10.6).

The matrix Y in the resolving pair, by virtue of (1.10.6) and (1.10.7), is a certain semi-inverse matrix to the matrix B, satisfying the conditions (1.10.6). As far as the matrix A^B is concerned, introducing it is then associated with the fact that it performs calculations and yields results, both of which are more compact.

For the subsequent discussion we must also note that, as a consequence of the equality (1.8.20), formula (1.10.5) for the matrix A^B can also be written as

$$A^B = Y(AY)^{\mathcal{D}}.$$

If Y is a certain fixed solution of the system (1.10.6), (1.10.7), then the matrix A^B is the only solution of the system

$$
\begin{align}
&(1)\ BA^BA = AA^BB, &&(1.10.9)\\
&\quad YAA^B = A^BAY, &&(1.10.10)\\
&\quad YBA^B = A^BBY, &&(1.10.11)\\
&(2)\ A^B = A^BAA^B, &&(1.10.12)\\
&(3)\ (YA)^{k+1}A^B = (YA)^kY, &&(1.10.13)\\
&\quad A^B(AY)^{s+1} = Y(AY)^s, &&(1.10.14)\\
&(4)\ A^BAYB = A^BA, &&(1.10.15)\\
&\quad BYAA^B = AA^B, &&(1.10.16)
\end{align}
$$

where k and s are, respectively, the indices of the matrices YA and AY.

The fact that the matrix $A^B = Y(AY)^{\mathcal{D}} = (YA)^{\mathcal{D}}Y$ satisfies the system (1.10.9)–(1.10.16) is verified by a simple substitution; therefore, it remains only to prove the uniqueness of the solution of this system.

Before beginning to prove it, we derive a number of useful corollaries from the system (1.10.9)–(1.10.16).

We multiply equation (1.10.10) from the left by the matrix B. Then, by virtue of (1.10.16), we obtain:

$$AA^B = BA^BAY. \tag{1.10.17}$$

Furthermore, by multiplying equation (1.10.11) from the left by the matrix A and also taking account of equation (1.10.9), we obtain:

$$AYBA^B = BA^BAY, \tag{1.10.18}$$

which shows the commutativeness of the matrices AY and BA^B.

In a similar way, by multiplying equations (1.10.10) and (1.10.11), respectively, by B and A from the right, we obtain:

$$A^BA = YAA^BB, \tag{1.10.19}$$

$$YAA^BB = A^BBYA \tag{1.10.20}$$

(the commutativeness of YA and A^BB).

Note also the equalities

$$A^B B = A^B B A^B A, \qquad B A^B = A A^B B A^B, \qquad (1.10.21)$$

$$(AA^B)^2 = AA^B, \qquad (A^B A)^2 = A^B A, \qquad (1.10.22)$$

which are readily obtainable with the help of equations (1.10.9) and (1.10.12).

Moreover, by multiplying the equality (1.10.10) by the matrix A from the left, we obtain:

$$AYAA^B = AA^B AY,$$

i.e. the matrices AY and AA^B are commutative and, hence,

$$AA^B(AY)^i = (AY)^i AA^B, \quad i = 0, 1, \ldots . \qquad (1.10.23)$$

In a similar way, we obtain:

$$A^B A(YA)^i = (YA)^i A^B A, \quad i = 0, 1, \ldots . \qquad (1.10.24)$$

Finally, by multiplying equation (1.10.13) by the matrix A from the right and by the matrix $(YA)^{\mathscr{D}}$ from the left, with the help of property (3) from (1.8.1), we arrive at the equality

$$(YA)^k(E - A^B A) = 0 \qquad (1.10.25)$$

(k is the index of the matrix YA).

Proceeding in exactly the same way, by multiplying equation (1.10.14) by A from the left and by $(AY)^{\mathscr{D}}$ from the right, we obtain:

$$(E - AA^B)(AY)^s = 0, \qquad (1.10.26)$$

where s is the index of the matrix AY.

Suppose now that the system (1.10.9)–(1.10.16) has two solutions, A_1^B and A_2^B, and $R = A_1^B - A_2^B$ is the difference between them. Then, for example, the equality (1.10.12) gives

$$A_1^B = A_1^B A A_1^B = R A A_1^B + A_2^B A A_1^B = R A A_1^B + A_2^B A R + A_2^B$$

and, consequently,

$$R = R A A_1^B + A_2^B A R. \qquad (1.10.27)$$

Moreover, from (1.10.13) and (1.10.14) the equalities

$$(YA)^{k+1} R = 0, \qquad R(AY)^{s+1} = 0 \qquad (1.10.28)$$

follow. From (1.10.27) and (1.10.28) it follows that $R = 0$. Indeed, as a consequence of (1.10.17), (1.10.18), (1.10.22) and (1.10.19), (1.10.20), (1.10.22) we have the equalities

$$AA_1^B = (AA_1^B)^{s+1} = (AYBA_1^B)^{s+1} = (AY)^{s+1}(BA_1^B)^{s+1},$$
$$A_2^B A = (A_2^B A)^{k+1} = (A_2^B BYA)^{k+1} = (A_2^B B)^{k+1}(YA)^{k+1},$$

by the use of which it is easy to show that the equalities (1.10.28) yield the equalities

$$RAA_1^B = 0, \qquad A_2^B AR = 0.$$

But then, according to (1.10.27), $R = 0$ and the uniqueness of the solution of the system (1.10.9)–(1.10.16) is thereby proved.

In the general case of an arbitrary pair of matrices (A, B), searching for a resolving pair of matrices presents problems. This is primarily attributable to the complexity of solving the system (1.10.6), (1.10.7). Fortunately, however, when one deals with the solution of degenerate systems of ordinary differential equations, the treatment should be confined only to those systems in which the pair of matrices (B, A) is perfect. In this case the solution of the system (1.10.6), (1.10.7) is simplified, and the construction of the matrix Y, satisfying the system (1.10.6), (1.10.7), becomes feasible. This construction will be presented in the next section.

Incidentally, we should note that if B is a nonsingular matrix, then $Y = B^{-1}$, $A^B = (B^{-1}A)^{\mathscr{D}}B^{-1}$, and the resolving pair for the pair of matrices (A, B) turns out to be unique. In particular, when $B = E$ we have $Y = E$ and $A^B = A^{\mathscr{D}}$.

If the elements of the matrices A and B are functions of the parameter $t \in [\alpha, \beta]$, then on the basis of Corollary 1.8.1 and formula (1.10.5), one can write the following equalities:

$$\begin{aligned} (A^B A)' &= (E - A^B A)((YA)^k)'((YA)^{\mathscr{D}})^k + ((YA)^{\mathscr{D}})^k((YA)^k)'(E - A^B A), \\ (AA^B)' &= (E - AA^B)((AY)^s)'((AY)^{\mathscr{D}})^s + ((AY)^{\mathscr{D}})^s((AY)^s)'(E - AA^B), \end{aligned} \qquad (1.10.29)$$

from which, in particular, it follows that

$$\begin{aligned} A^B A(A^B A)'(A^B A) &= 0, \quad AA^B(AA^B)'AA^B = 0, \\ (A^B A)' &= (A^B A)'A^B A + A^B A(A^B A)', \qquad\qquad (1.10.30) \\ (AA^B)' &= (AA^B)'AA^B + AA^B(AA^B)' \end{aligned}$$

(when deriving (1.10.30) it is necessary to take (1.10.22) into account).

Let us now consider the system

$$A(\lambda x) = Bx + f, \qquad (1.10.31)$$

as well as assuming that the pair of matrices in it is perfect. (The transition from an arbitrary system (1.10.31) to a system with a perfect pair of matrices (B, A) can be performed with the help of Theorem 1.6.2; according to this theorem the pair of matrices $(B - \alpha A, A)$ in the system $A[(\lambda - \alpha)x] = (B - \alpha A)x + f$ (obviously, equivalent to the system (1.10.31)), for almost all α, is perfect.)

Let x be the solution of the system (1.10.31). Then, by multiplying the equality (1.10.31) by the matrix Y, we obtain:

$$\lambda Y Ax = x - u + Yf, \qquad (1.10.32)$$

where

$$u = (E - YB)x \qquad (1.10.33)$$

and, consequently,

$$Bu = 0. \tag{1.10.34}$$

Furthermore, as in the case of equation (1.10.1), by iterating (1.10.32) we arrive at the equality

$$x = \lambda^k (YA)^k - [\lambda^{k-1}(YA)^{k-1} + \cdots + E](Yf - u).$$

If we multiply this by the matrix $(E - A^B A)$, as a consequence of (1.10.24) and (1.10.25), we obtain:

$$(E - A^B A)x = -(E - A^B A) \sum_{i=0}^{k-1} \lambda^i (YA)^i (Yf - u),$$

which, in view of (1.10.15), (1.10.24), and (1.10.33), can also be written as

$$(E - A^B A)x = -(E - A^B A) \sum_{i=0}^{k-1} \lambda^i (YA)^i Yf + \sum_{i=0}^{k-1} \lambda^i (YA)^i u. \tag{1.10.35}$$

We now multiply the equality (1.10.31) by the matrix A^B. As a result, taking into account the first equality from (1.10.21), we obtain:

$$\lambda(A^B A x) = A^B B (A^B A x) + A^B f,$$

from which it follows that if λ is not an eigen-number of the matrix $A^B B$, then

$$A^B A x = (\lambda E - A^B B)^{-1} A^B f. \tag{1.10.36}$$

By adding (10.35) to (1.10.36) we obtain:

$$x = (\lambda E - A^B B)^{-1} A^B f - (E - A^B A) \sum_{i=0}^{k-1} \lambda^i (YA)^i Yf + \sum_{i=0}^{k-1} \lambda^i (YA)^i u. \tag{1.10.37}$$

Finally, by multiplying equation (1.10.31) by the matrix $E - BY$, we arrive at the equality

$$\lambda(E - BY)Ax = (E - BY)f, \tag{1.10.38}$$

which must be satisfied by the vector (1.10.37) and, therefore, substitution of the vector (1.10.37) into the equality (1.10.38) must lead to the simultaneity condition for the system (1.10.31).

On substituting (1.10.37) into (1.10.38), in this case taking into account the perfectness of the pair of matrices (B, A), i.e. the validity of the equalities

$$(E - BY)A(YA)^i (E - YB) = 0, \quad i = 0, 1, \ldots, \tag{1.10.39}$$

as well as the properties (1.10.8) and (1.10.16) and the equalities (1.10.33) and

$$(\lambda E - A^B B)^{-1} A^B = A^B (\lambda E - BA^B)^{-1},$$

we obtain:

$$\sum_{i=0}^{r} \lambda^i (E - BY)(AY)^i f = 0, \tag{1.10.40}$$

where $r = \min(k, s - 1)$, k is the index of the matrix YA, and s is the index of the matrix AY.

If the simultaneity condition (1.10.40) is satisfied, then, using the properties of the resolving pair of matrices (A^B, Y) as well as the equalities (1.10.39), using a direct substitution one can show that for any vector u satisfying equation (1.10.34), the vector (1.10.37) satisfies equation (1.10.31). Thus, the equality (1.10.40) is a necessary and sufficient condition for solvability of the system (1.10.31), and formula (1.10.37), in which u is an arbitrary solution of equation (1.10.34), is a general solution of this system.

Let us now consider the homogeneous equation

$$A(\lambda x) = Bx \qquad (1.10.41)$$

in order to ascertain the conditions under which this equation has only a zero solution or, in other words, the equality $\text{def}(B - \lambda A) = 0$ is satisfied for its pair of matrices (A, B).

As before, we assume that the pair of matrices (B, A) is perfect. Then, if λ is not an eigen-number of the matrix $A^B B$, the general solution of equation (1.10.41), according to formula (1.10.37) (in view of the fact that $u = (E - YB)v$, where v is an arbitrary vector) has the form

$$x = \sum_{i=0}^{k-1} \lambda^i (YA)^i (E - YB)v,$$

from which, owing to the arbitrariness of v, it follows that equation (1.10.41) has only a zero solution if and only if

$$\sum_{i=0}^{k-1} \lambda^i (YA)^i (E - YB) = 0. \qquad (1.10.42)$$

Now, one can prove the following theorem.

Theorem 1.10.1 If the pair of matrices (B, A) is perfect and λ is not an eigen-number of the matrix $A^B B$, then the equality

$$\text{def}(B - \lambda A) = 0$$

is valid if and only if $E - YB = 0$ (in other words, if the columns of the matrix B are linearly independent (see Corollary 1.4.13)).

Proof If the equaltiy (1.10.42) is satisfied, then by multiplying it by the matrix $\lambda^{k-1}(YA)^{k-1}$ from the left, with the help of the second equality from (1.10.8), we obtain $\lambda^{k-1}(YA)^{k-1}(E - YB) = 0$. Taking into account this result and multiplying the equality (1.10.42) by the matrix $\lambda^{k-2}(YA)^{k-2}$, we have $\lambda^{k-2}(YA)^{k-2}(E - YB) = 0$, etc. Ultimately, we arrive at the equality $E - YB = 0$, which means that the columns of the matrix B are linearly independent. If, however, $E - YB = 0$ (the columns of the matrix B are linearly independent), then obviously the equality (1.10.42) is valid. The theorem is thereby proved.

1.11. An algorithm for obtaining a resolving pair of matrices

The main problem when determining the resolving pair of matrices (A^B, Y), corresponding to the pair of matrices (A, B), is that of solving the system of equations (1.10.6), (1.10.7) for the matrix Y (the matrix A^B is obtained from the formula (1.10.5)). In this section the solution of the system (1.10.6), (1.10.7) will be constructed under the assumption that the pair of matrices (B, A) is perfect. From the subsequent treatment it will become clear that such an assumption is not a limiting one (by considering an algebraic system; this has been demonstrated in the previous section).

Before starting construction, we note that the system (1.10.6), (1.10.7), which is satisfied by the desired matrix Y, is equivalent to the system

$$(E - BB^-)(AY)^{\mathscr{D}} = 0, \qquad (YA)^{\mathscr{D}}(E - B^- B) = 0, \qquad (1.11.1)$$

$$BYB = B, \qquad (1.11.12)$$

for any semi-inverse matrix B^- to the matrix B. Indeed, by multiplying the first equation in (1.11.1) by the matrix $E - BY$ from the left, and the second by the matrix $(E - YB)$ from the right, we obtain equations (1.10.6). The opposite is proved in a similar manner: it is necessary to multiply the first equation in (1.10.6) by the matrix $E - BB^-$ from the left, and the second by the matrix $E - B^- B$ from the right.

Let us now formulate and prove several simple lemmas.

Lemma 1.11.1 Let the sequence of matrices A_i $(i = 0, 1, \ldots)$ be obtained from the formulae $A_{i+1} = A_i K_i$, where $A_0 = A$ and K_i is a given matrix. Then, if the matrix Y satisfies the first k equations of the system

$$(E - A_{i+1} A_{i+1}^-) A_i A_i^- A Y A_i = 0, \quad i = 0, 1, \ldots, \qquad (1.11.3)$$

then it satisfies also the first k equations of the system

$$(E - A_{i+1} A_{i+1}^-) A Y A_i = 0, \quad i = 0, 1, \ldots. \qquad (1.11.4)$$

Proof Since $A_0 = A$, then $A_0 A_0^- A = A$, and the first equations of the systems (1.11.3) and (1.11.4) coincide. From this, the validity of the lemma for $k = 1$ follows.

Suppose that the lemma is true at a certain $k = 1$, and the matrix Y satisfies the first $k + 1$ equations of the system (1.11.3). Then, by virtue of the assumption, the matrix Y also satisfies the first k equations of the system (1.11.4) and, consequently, with respect to the $(k + 1)$th equation of the system (1.11.4), we can write

$$\begin{aligned}
(E - A_{k+1} A_{k+1}^-) A Y A_k &= (E - A_{k+1} A_{k+1}^-) A Y A_{k-1} K_{k-1} \\
&= (E - A_{k+1} A_{k+1}^-) A_k A_k^- A Y A_{k-1} K_{k-1} \\
&= (E - A_{k+1} A_{k+1}^-) A_k A_k^- Y A_k = 0.
\end{aligned}$$

Thus, Lemma 1.11.1 is proved by induction in k.

Lemma 1.11.2 If the matrix Y satisfies the first k equations of the system (1.11.4), then

$$(E - A_k A_k^-)(AY)^k A = 0. \tag{1.11.5}$$

Proof If kY satisfies the first k equations of the system (1.11.4), then, in particular, $(E - A_k A_k^-)AYA_{k-1} = 0$. By multiplying this equality from the right by the matrix

$$A_{k-1}^-(AY)A_{k-2}A_{k-2}^-(AY)\dots A_1 A_1^-(AY)A_0,$$

taking into account the designation $A_0 = A$ and all of the previous equations of the system (1.11.4), we obtain (1.11.5). Lemma 1.11.2 is thereby proved.

Lemma 1.11.3 If the matrix Y satisfies equation (1.11.5) and the equality

$$(E - BB^-)A_k = 0 \tag{1.11.6}$$

holds, then

$$(E - BB^-)(AY)^k A = 0. \tag{1.11.7}$$

Proof It is sufficient to multiply the equality (1.11.5) from the left by the matrix $(E - BB^-)$, in this case taking into account the equality (1.11.6).

Lemma 1.11.4 If the matrix Y is the solution of equation (1.11.7), then

$$(E - BB^-)(AY)^{\mathscr{D}} = 0. \tag{1.11.8}$$

Proof On the basis of Theorem 1.8.5,

$$(AY)^{\mathscr{D}} = (AY)^r [(AY)^{2r+1}]^-(AY)^r$$

at any $r + 1$ not smaller than the index of the matrix AY. But then, by multiplying the equality (1.11.7) from the right by the matrix

$$Y(AY)^{r-1}[(AY)^{2(k+r)+1}]^-(AY)^{k+r},$$

we obtain (1.11.8).

Lemma 1.11.5 If the pair of matrices (B, A) is perfect, and the matrix Y is the solution of the system

$$(E - BB^-)(AY)^k A = 0, \quad BYB = B, \tag{1.11.9}$$

then the system (1.11.9) is also satisfied by all matrices of the form

$$Z = Y + (E - B^-B)U,$$

where U is an arbitrary matrix.

Proof Note at first that any semi-inverse matrix B^-, rather than only the matrix Y, can be substituted into the extreme co-factors of the left-hand side of the equalities (1.10.39) which define the perfect pair of matrices (B, A). This can be demonstrated by reasoning as was done when proving the equivalency of the systems (1.11.1), (1.11.2) and (1.10.6), (1.10.7). Taking this into consideration, instead of the equalities (1.10.39) we shall use the equalities

$$(E - BB^-)A(YA)^i(E - B^-B) = 0.$$

In this case the assertion of Lemma 1.11.5 becomes the corollary of the equalities

$$
\begin{aligned}
(E - BB^-)(AZ)^k A &= (E - BB^-)[AY + A(E - B^-B)](AZ)^{k-1}A \\
&= (E - BB^-)(AY)(AZ)^{k-1}A \\
&= (E - BB^-)[(AY)^2 + (AY)A(E - B^-B)](AZ)^{k-2}A \\
&= (E - BB^-)(AY)^2(AZ)^{k-2}A = \cdots = (E - BB^-)(AY)^k A = 0
\end{aligned}
$$

(it is clear that the matrix Z also satisfies the second equation from (1.11.9) because it is a semi-inverse matrix of the matrix B).

The proofs of the following five lemmas are performed by analogy with the proofs of Lemmas 1.11.1–1.11.5.

Lemma 1.11.6 Let the sequence of matrices $\tilde{A}_i$ ($i = 0, 1, \ldots$) be obtainable from the formulae $\tilde{A}_{i+1} = \tilde{K}_i \tilde{A}$, where $\tilde{A}_0 = A$ and $\tilde{K}_i$ are given matrices. Then, if the matrix $\tilde{Y}$ satisfies the first s equations of the system

$$\tilde{A}_i \tilde{Y} A \tilde{A}_i^- \tilde{A}_i(E - \tilde{A}_{i+1}^- \tilde{A}_{i+1}) = 0, \quad i = 0, 1, \ldots, \tag{1.11.10}$$

then it also satisfies the first s equations of the system

$$\tilde{A}_i \tilde{Y} A(E - \tilde{A}_{i+1}^- \tilde{A}_{i+1}) = 0, \quad i = 0, 1, \ldots. \tag{1.11.11}$$

Lemma 1.11.7 If the matrix $\tilde{Y}$ satisfies the first s equations of the system (1.11.11), then

$$A(\tilde{Y}A)^s(E - \tilde{A}_s^- \tilde{A}_s) = 0. \tag{1.11.12}$$

Lemma 1.11.8 If the matrix $\tilde{Y}$ satisfies equation (1.11.12) and

$$\tilde{A}_s(E - B^-B) = 0, \tag{1.11.13}$$

then

$$A(\tilde{Y}A)^s(E - B^-B) = 0. \tag{1.11.14}$$

Lemma 1.11.9 If the matrix $\tilde{Y}$ is the solution of equation (1.11.14), then

$$(\tilde{Y}A)^{\mathcal{P}}(E - B^-B) = 0. \tag{1.11.15}$$

Lemma 1.11.10 If the pair of matrices (B, A) is perfect, and the matrix $\tilde{Y}$ is the solution of the system

$$A(\tilde{Y}A)^s(E - B^-B) = 0, \qquad B\tilde{Y}B = B, \tag{1.11.16}$$

then the system (1.11.16) is also satisfied by all matrices of the form

$$\tilde{Z} = \tilde{Y} + V(E - BB^-),$$

where V is an arbitrary matrix.

Lemma 1.11.11   If the pair of matrices (B, A) is perfect, and the matrices are of the form

$$Y = B^- + V(E - BB^-) \tag{1.11.17}$$

and

$$\tilde{Y} = B^- + (E - B^- B)U \tag{1.11.18}$$

and satisfy, respectively, the systems (1.11.9) and (1.11.16), then the matrix

$$Z = B^- + V(E - BB^-) + (E - B^- B)U$$

is the solution (for Y) of the system

$$(E - BB^-)(AY)^k = 0, \qquad A(YA)^s(E - B^- B) = 0, \tag{1.11.19}$$

$$BYB = BB. \tag{1.11.20}$$

Proof   Since, according to (1.11.17) and (1.11.18) the matrix Z has two representations

$$Z = Y + (E - B^- B)U = \tilde{Y} + V(E - BB^-),$$

then, according to Lemma 1.11.5, it satisfies the system (1.11.9), and according to lemma 1.11.10 it satisfies the system (1.11.16). From this it easily follows that the matrix Z also satisfies the system (1.11.19), (1.11.20).

Finally, from lemmas 1.11.4 and 1.11.9 follows yet another lemma.

Lemma 1.11.12   If the matrix Y is the solution of the system (1.11.19), (1.11.20), then it satisfies the system (1.11.1), (1.11.2) and, consequently, the system (1.10.6), (1.10.7) of interest to us:

$$(E - BY)(AY)^{\mathscr{D}} = 0, \qquad (YA)^{\mathscr{D}}(E - YB) = 0, \tag{1.11.21}$$

$$BYB = B. \tag{1.11.22}$$

From this chain of lemmas following from each other, we have:

Theorem 1.11.1   Let the sequences of matrices $A_i \, (i = 0, 1, \ldots)$ and $\tilde{A}_i \, (i = 0, 1, \ldots)$ be constructed by the formulae

$$A_{i+1} = A_i K_i, \quad \tilde{A}_{i+1} = \tilde{K}_i \tilde{A}_i, \qquad A_0 = \tilde{A}_0 = A, \quad i = 0, 1, \ldots, \tag{1.11.23}$$

and the matrices K_i and $\tilde{K}_i$ be such that there exist such non-negative integer numbers k and s that

$$(E - BB^-)A_k = 0, \qquad \tilde{A}_s(E - B^- B) = 0 \qquad (1.11.24)$$

(see (1.11.6) and (1.11.13)). Then, if there exist matrices V and U satisfying, respectively, the systems

$$(E - A_{i+1}A_{i+1}^-)A_iA_i^- A[B^- A_i + V(A_i - BB^- A_i)] = 0,$$
$$i = 0,\ldots,k-1, \quad (1.11.25)$$

and

$$[\tilde{A}_i B^- + (\tilde{A}_i - \tilde{A}_i B^- B)U]A\tilde{A}_i^- \tilde{A}_i(E - \tilde{A}_{i+1}^- \tilde{A}_{i+1}) = 0,$$
$$i = 0,\ldots,s-1, \quad (1.11.26)$$

then the solution of the system (1.11.21), (1.11.22) also exists and is expressed by the formula

$$Y = B^- + V(E - BB^-) + (E - B^- B)U. \qquad (1.11.27)$$

Thus, if the sequences (1.11.23) with properties (1.11.24) are found and the solutions of the systems (1.11.25), (1.11.26) exist, then the solution of the system (1.11.21), (1.11.22) is reduced to solving the systems (1.11.25), (1.11.26).

The form of the required sequences can be obtained by representing the solutions of the systems (1.11.25), (1.11.26) with the help of semi-inverse matrices. In this case it appears that for the sequence $A_i (i = 0, 1,\ldots)$ one can take the sequence

$$A_{i+1} = A_i K_i, \quad i = 0, 1,\ldots, \qquad A_0 = A, \qquad (1.11.28)$$

where

$$K_i = A_i^- AB^- A_i R_i,$$
$$R_i = E - (A_i - BB^- A_i)^-(A_i - BB^- A_i),$$

and for the sequence $\tilde{A}_i (i = 0, 1,\ldots)$ one can take the sequence

$$\tilde{A}_{i+1} = \tilde{K}_i \tilde{A}_i, \quad i = 0, 1,\ldots, \qquad \tilde{A}_0 = A, \qquad (1.11.29)$$

where

$$\tilde{K}_i = \tilde{R}_i \tilde{A}_i B^- A \tilde{A}_i^-,$$
$$\tilde{R}_i = E - (\tilde{A}_i - \tilde{A}_i B^- B)(\tilde{A}_i - \tilde{A}_i B^- B)^-.$$

First of all we prove that the sequence (1.11.28) has the property of (1.11.24). For this purpose we note that the known theorem of the rank of the product of matrices leads to the inequalities

$$\text{rank } A_{i+1} \leqslant \text{rank } A_i, \qquad (1.11.30)$$

$$\text{rank } A_{i+1} \leqslant \text{rank } A_i R_i, \qquad (1.11.31)$$

$$\text{rank } A_i R_i \leqslant \text{rank } A_i, \quad i = 0, 1,\ldots. \qquad (1.11.32)$$

Let k be the first of the non-negative integer numbers for which rank A_{k+1} = rank A_k. Such a k will, of course, be found because the rank of any matrix is a non-negative integer number and, consequently, an unlimited rigorous decrease of the ranks in (1.11.30) is impossible. With such a k, it follows from (1.11.31) and (1.11.32) that rank A_k = rank $A_k R_k$ and, therefore, on the basis of Lemma 1.4.9 we have

$$\text{rank}\,[(E - BB^-)A_k] = \text{rank}\,A_k - \text{rank}(A_k R_k) = 0,$$

i.e. $(E - BB^-)A_k = 0$. Hence, the existence of the desired k is thereby proved and, consequently, the chain (1.11.28) has the property of (1.11.24). The same result occurs also with respect to the chain (1.11.29) (the proof is similar).

While beginning to solve the system (1.11.25), we introduce the designations

$$
\begin{aligned}
P_i &= (E - BB^-)A_i, & S_{i+1} &= E - A_{i+1}A_{i+1}^-, \\
M_{i+1} &= S_{i+1}A_iA_1^- A, & R_i &= E - P_i^- P_i.
\end{aligned}
\tag{1.11.33}
$$

Then the system (1.11.25) will be written as

$$M_{i+1}VP_i = -M_{i+1}B^- A_i, \qquad i = 0, 1, \ldots, k-1. \tag{1.11.34}$$

Our purpose is to ascertain and to show that, if the sequence of matrices $V_i (i = 0, 1, \ldots)$ is obtainable from the formulae

$$
\begin{aligned}
V_{i+1} &= V_i + G_i(M_{i+1}G_i)^- M_{i+1}(B^- A_i - V_i A_i)P_i^- (E - BB^-), \\
G_{i+1} &= G_i[E - (M_{i+1}G_i)^- (M_{i+1}G_i)], \quad i = 0, 1, \ldots, \\
A_0 &= A, \quad G_0 = E, \quad V_0 = 0,
\end{aligned}
\tag{1.11.35}
$$

then V_k, where k is an integer number, for which $(E - BB^-)A_k \equiv P_k = 0$, satisfies the system (1.11.34).

In order to make sure that this is true, we first prove several lemmas.

Lemma 1.11.13   The equalities

$$M_{i+1}B^- A_i P_i^- P_i = M_{i+1}B^- A_i, \quad i = 0, 1, \ldots$$

are valid.

Proof   Taking into consideration the notation of (1.11.33) and the obvious equality $\hat{S}_{i+1}A_{i+1} = 0$, we obtain:

$$M_{i+1}B^- A_i(E - P_i^- P_i) = M_{i+1}B^- A_i R_i = S_{i+1}A_iA_i^- AB^- A_iR_i = S_{i+1}A_{i+1} = 0.$$

Lemma 1.11.14   The equalities

$$M_iG_k = 0, \quad i = 1, \ldots, k$$

are valid (at any k).

Proof   When $k = 1$, the assertion of the lemma follows immediately from the formula for G_1 (see (1.11.35)) and from the determining property $NN^-N = N$

for semi-inverse matrices. If, however, the lemma is true for a certain $k \geqslant 1$, then, by virtue of the fact that

$$G_{k+1} = G_k[E - (M_{k+1}G_k)^-(M_{k+1}G_k)]$$

(see (1.11.35)) for $i = 1, \ldots, k$, we have $M_i G_{k+1} = 0$, and for $i = k + 1$ we obtain the equality $M_{k+1}G_{k+1} = 0$, which follows from the determining property of the semi-inverse matrices.

Lemma 1.11.15 For all $i = 0, 1, \ldots$, the equalities $V_i B = 0$ hold.

Proof By multiplying from the right the recurrent formula (1.11.35) for V_i by the matrix B, we obtain $V_{i+1}B = V_i B$. But $V_0 = 0$. From this follows the assertion of the lemma.

Corollary 1.11.1 $V_i A_i P_i^- P_i = V_i A_i$, $\quad i = 0, 1, \ldots$.

Proof By virtue of Lemma 1.11.15,

$$V_i A_i = V_i(E - BB^-)A_i = V_i P_i.$$

But then

$$V_i A_i P_i^- P_i = V_i P_i P_i^- P_i = V_i P_i = V_i A_i.$$

Lemma 1.11.16 $B^- A_i - V_i A_i = G_i B^- A_i$, $\quad i = 0, 1, \ldots$.

Proof When $i = 0$, the lemma is obvious because

$$V_0 = 0, \qquad G_0 = E.$$

Let the lemma be true for a certain $i \geqslant 0$. Then, by making use of the notations in (1.11.28), (1.11.33), and (1.11.35), Lemma 1.11.13, and the corollary from Lemma 1.11.15, we obtain:

$$
\begin{aligned}
& B^- A_{i+1} - V_{i+1}A_{i+1} \\
&= [B^- A_i - V_i A_i - (V_{i+1} - V_i)A_i]A_i^- AB^- A_i R_i \\
&= [G_i B^- A_i - G_i(M_{i+1}G_i)^- M_{i+1}(B^- A_i - V_i A_i)P_i^- P_i]A_i^- AB^- A_i R_i \\
&= [G_i B^- A_i - G_i(M_{i+1}G_i)^- M_{i+1}(B^- A_i - V_i A_i)]A_i^- AB^- A_i R_i \\
&= [G_i B^- A_{i+1} - G_i(M_{i+1}G_i)^- M_{i+1}G_i B^- A_{i+1}] \\
&= G_i[E - (M_{i+1}G_i)^- M_{i+1}G_i]B^- A_{i+1} = G_{i+1}B^- A_{i+1}.
\end{aligned}
$$

Thus, Lemma 1.11.16 is proved by induction.

Lemma 1.11.17 The matrix V_s for $s \geqslant 1$ satisfies the first s equations of the system

$$M_{i+1}VP_i = -M_{i+1}B^- A_i. \tag{1.11.36}$$

Proof   When $s = 1$, the lemma is true because $V_0 = 0$, $G_0 = E$, and, by virtue of Lemmas 1.11.15 and 1.11.13, as well as by formula (1.11.35), we have

$$
\begin{aligned}
- M_1 V_1 P_0 &= - M_1 V_1 (E - BB^-) A_0 = - M_1 V_1 A_0 \\
&= - M_1 M_1^- M_1 B^- A_0 P_0^- P_0 = - M_1 B^- A_0 P_0^- P_0 = - M_1 B^- A_0.
\end{aligned}
$$

Let the lemma be true at a certain $s \geqslant 1$. Then, on the basis of the inductive assumption, formula (1.11.35), and of Lemma 1.11.14, when $i = 0, \ldots, s - 1$, we can write

$$
\begin{aligned}
- M_{i+1} V_{s+1} P_i &= - M_{i+1} [V_s + G_s (M_{s+1} G_s)^- M_{s+1} (B^- A_s - V_s A_s) P_s^-] P_i \\
&= - M_{i+1} V_s P_i = - M_{i+1} B^- A_i,
\end{aligned}
$$

and when $i = s$, taking into account Lemmas 1.11.13 and 1.11.16 as well as Lemma 1.11.15 and the corollary from it, we have

$$
\begin{aligned}
- M_{s+1} V_{s+1} P_s &= - M_{s+1} V_{s+1} A_s \\
&= - M_{s+1} [V_s A_s + G_s (M_{s+1} G_s)^- M_{s+1} (B^- A_s - V_s A_s) P_s^- P_s] \\
&= - M_{s+1} V_s A_s - M_{s+1} G_s (M_{s+1} G_s)^- M_{s+1} (B^- A_s - V_s A_s) \\
&= - M_{s+1} V_s A_s - M_{s+1} G_s (M_{s+1} G_s)^- M_{s+1} G_s B^- A_s \\
&= - M_{s+1} V_s A_s - M_{s+1} G_s B^- A_s = - M_{s+1} [V_s A_s + G_s B^- A_s] \\
&= - M_{s+1} B^- A_s.
\end{aligned}
$$

Thus, the matrix V_{s+1} satisfies the first $s + 1$ equations of the system (1.11.36). Induction is carried out, and Lemma 1.11.17 is proved.

By applying this lemma to the matrix V_k, where k is a number satisfying the condition of Theorem 1.11.1 (see (1.11.24)), we find that $- V_k$ is the solution of the system (1.11.25).

Proceeding in a similar manner, using the sequence $\tilde{A}_i$ $(i = 0, 1, \ldots)$, it is possible to construct the solution of the system (1.11.26) (the matrix $- U_s$) and, as a result, on the basis of Theorem 1.11.1, to obtain the solution of (1.11.27) of the system (1.11.21), (1.11.22) of interest:

$$
Y = B^- - V_k (E - BB^-) - (E - B^- B) U_s. \tag{1.11.37}
$$

In this case it should be noted that, as a consequence of the equalities $V_k B = 0$ and $BU_s = 0$ (see Lemma 1.11.15), formula (1.11.37) can be written as

$$
Y = B^- - V_k - U_s. \tag{1.11.38}
$$

Thus, if the pair of matrices (B, A) is perfect, then the solution algorithm for the system (1.11.21), (1.11.22) implies the following.

(1) With the help of the recurrent formula

$$
\begin{aligned}
A_{i+1} &= A_i A_i^- AB^- A_i [E - (A_i - BB^- A_i)^- (A_i - BB^- A_i)], \\
A_0 &= A, \quad i = 0, 1, \ldots,
\end{aligned}
$$

we define the matrices A_i until, at a certain $i = k$, the equality $(E - BB^-)A_k = 0$ is obtained.

(2) At the same time, by the formulae

$$\tilde{A}_{i+1} = [E - (\tilde{A}_i - \tilde{A}_i B^- B)(\tilde{A}_i - \tilde{A}_i B^- B)^-]\tilde{A}_i B^- A\tilde{A}_i^- \tilde{A}_i,$$
$$\tilde{A}_0 = A, \quad i = 0, 1, \ldots,$$

we calculate the matrices $\tilde{A}_i$ up to the matrix $\tilde{A}_s$, at which $\tilde{A}_s(E - B^- B) = 0$.

(3) By the recurrent formulae

$$M_{i+1} = (E - A_{i+1}A_{i+1}^-)A_i A_i^- A,$$
$$P_i = (E - BB^-)A_i,$$
$$V_{i+1} = V_i + G_i(M_{i+1}G_i)^- M_{i+1}(B^- A_i - V_i A_i)P_i^-(E - BB^-),$$
$$G_{i+1} = G_i[E - (M_{i+1}G_i)^-(M_{i+1}G_i)],$$
$$A_0 = A, \quad G_0 = E, \quad V_0 = 0, \quad i = 0, 1, \ldots, k - 1,$$

we determine the matrix V_k.

(4) At the same time, by the formulae

$$\tilde{M}_{i+1} = A\tilde{A}_i^- \tilde{A}_i(E - \tilde{A}_{i+1}^- \tilde{A}_{i+1}),$$
$$\tilde{P}_i = \tilde{A}_i(E - B^- B),$$
$$U_{i+1} = U_i + (E - B^- B)\tilde{P}_i^-(\tilde{A}_i B^- - \tilde{A}_i U_i)\tilde{M}_{i+1}(\tilde{G}_i \tilde{M}_{i+1})^- \tilde{G}_i,$$
$$\tilde{G}_{i+1} = [E - (\tilde{G}_i \tilde{M}_{i+1})(\tilde{G}_i \tilde{M}_{i+1})^-]\tilde{G}_i,$$
$$\tilde{A}_0 = A, \quad \tilde{G}_0 = E, \quad U_0 = 0, \quad i = 0, 1, \ldots, s - 1,$$

we determine the matrix U_s.

(5) By formula (1.11.38) we obtain the solution of the system (1.11.21), (1.11.22).

Remark 1.11.1 If the equality $(E - BB^-)A_0 = 0$ is satisfied, then $V_0 = 0$ in formula (1.11.38). In exactly the same way, if $\tilde{A}_0(E - B^- B) = 0$, then $U_0 = 0$ in (1.11.38). Moreover we note that from the equalities $(E - BB^-)A_k = 0$ and $\tilde{A}_s(E - B^- B) = 0$ the equalities $(E - BB^-)A_i = 0$ (when $i > k$) and $\tilde{A}_i(E - B^- B) = 0$ (when $i > s$) follow (we leave the proof for the reader; hence, it is useful to see [14–16]).

2 DIFFERENCE SCHEMES FOR SOLVING SINGULAR SYSTEMS OF ORDINARY DIFFERENTIAL EQUATIONS

In this chapter we construct and substantiate several difference schemes that are applied to solve Cauchy problems related to singular systems of ordinary differential equations. In this case it will be assumed throughout that the matrices and vectors involved in differential and difference equations (including generalized inverse matrices) are sufficiently smooth ones, and there exists a unique sufficiently smooth solution of the initial differential problem. For the sake of simplicity, when constructing the schemes we have adopted the implicit scheme of the first order of accuracy. However, from the presentation it should become clear that schemes of increased order of accuracy can be obtained by proceeding along the same vein.

2.1. A difference scheme for solving systems with a perfect group of three matrices

A general theory of singular problems with a perfect group of three matrices (A, B, C) will be presented in the next chapter; for the time being, we note that the simplest representation of such problems is provided by the problem

$$A(t)x'(t) = B(t)x(t) + f(t), \quad \alpha \leqslant t \leqslant \beta,$$

$$Cx(\alpha) = a,$$

in which the group of three matrices $(A(t), B(t), C)$ is perfect (see Definition 1.7.1).

Since at this point our intent is only to examine the Cauchy problem $(C = E)$, then, according to Corollary 1.7.1, it must be assumed that the columns of the matrix $A(t)$ for all $t \in [\alpha, \beta]$ are linearly independent.

Thus, let us consider the problem

$$A(t)x'(t) = B(t)x(t) + f(t), \quad \alpha \leqslant t \leqslant \beta, \tag{2.1.1}$$

$$x(\alpha) = a, \tag{2.1.2}$$

where the columns of the matrix $A(t)$ are linearly independent.

By multiplying equation (2.1.1) by the matrix A^+, taking into account the fact that $A^+ A = E$ (see Lemma 1.4.13), we obtain:

$$x' = A^+ Bx + A^+ f, \tag{2.1.3}$$

$$x(\alpha) = a \tag{2.1.4}$$

(for brevity, we omit mentioning the dependence on t).

From the general assumption adopted in this chapter regarding the uniqueness of the solutions of the problems concerned and the uniqueness of the solution of the Cauchy problems (2.1.3) and (2.1.4), it follows that problems (2.1.1), (2.1.2) and (2.1.3), (2.1.4) are equivalent. It should, however, be noted that the assumption about the uniqueness of the solution of the problem (2.1.1), (2.1.2) is actually superfluous: in the next chapter it will be shown that, if the solution x of the problem (2.1.1), (2.1.2) with the matrix A, satisfying the condition $A^+ A = E$, exists, then it is unique. Without going, for the time being, into detail associated with the formulation of the theorems about the solution existence, we only note that, if equation (2.1.1) is multiplied by the matrix $E - AA^+$, then the equation

$$(E - AA^+)Bx = -(E - AA^+)f$$

is obtained, from which, when $t = \alpha$, follows the condition for matching the initial data with the right-hand side of

$$[E - A(\alpha)A^+(\alpha)]B(\alpha)a = -[E - A(\alpha)A^+(\alpha)]f(\alpha).$$

To solve the problem (1.2.3), (1.2.4), which is equivalent to the initial problem (1.2.1), (1.2.2), we use the following implicit difference scheme:

$$\frac{x_i - x_{i-1}}{\tau} = A_i^+ B_i x_i + A_i^+ f_i, \quad i = 1, \ldots, N; \qquad N\tau = \beta - \alpha, \tag{2.1.5}$$

$$x_0 = a, \tag{2.1.6}$$

where (as in the subsequent treatment)

$$A_i^+ = A^+(i\tau), \qquad B_i = B(i\tau), \qquad f_i = f(i\tau),$$

and x_i, as will be shown below, is the value of an approximate solution of the problem (2.1.3), (2.1.4) (or, which is the same, of the problem (2.1.1), (2.1.2)) at the point $t = i\tau$.

The difference scheme (2.1.5), (2.1.6) is well studied; however, in order not to repeat in the subsequent treatment, we shall, nevertheless, prove that its solution x_i when $\tau \to 0$ tends to the value of an accurate solution $x(i\tau)$. For this purpose we use the procedure known in the theory of difference schemes: let $v_i = x_i - x(i\tau)$, then, after substitution of $x_i = v_i + x(i\tau)$ into the difference equation (2.1.5) for deviations v_i, we obtain:

$$v_i = (E - \tau A_i^+ B_i)^{-1} v_{i-1} + (E - \tau A_i^+ B_i)^{-1} [- (x(i\tau) - x((i-1)\tau))$$
$$+ \tau A_i^+ B_i x(i\tau) + \tau f(i\tau)], \tag{2.1.7}$$

$$v_0 = 0, \quad i = 1, \dots, N, \qquad N\tau = \beta - \alpha.$$

Furthermore, we make use of the formula

$$x((i-1)\tau) = x(i\tau) - \tau x'(i\tau) + \frac{\tau^2}{2} x''(i\theta\tau), \tag{2.1.8}$$

where $\theta \in [0, 1]$ and is assumed dependent on the number of components of the vector x''.

Upon substituting (2.1.8) into (2.1.7) we obtain the formula

$$v_i = C_i v_{i-1} + (\tau^2/2)\varphi_i,$$

in which $C_i = (E - \tau A_i^+ B_i)^{-1}$ and $\varphi_i = C_i x''(i\theta\tau)$.

As a consequence of the sufficient smoothness of the matrices A^+ and B and of the vector x'', there obviously exist constant K and L independent of $i = 1, \dots, N$ and of τ (from a certain interval $(0, \tau_0)$) such that

$$\| C_i \| \leqslant 1 + K\tau, \qquad \| \varphi_i \| \leqslant L.$$

But then

$$\| v_i \| \leqslant (1 + K\tau) \| v_{i-1} \| + \tau^2 L, \qquad \| v_0 \| = 0,$$

and, consequently,

$$\| v_i \| \leqslant L\tau^2 \sum_{s=0}^{i-1} (1 + K\tau)^s.$$

We now note that since $s\tau \leqslant \beta - \alpha$ and $i\tau \leqslant \beta - \alpha$, then the inequalities

$$(1 + K\tau)^s = (1 + K\tau)^{Ks\tau/K\tau} \leqslant e^{K(\beta - \alpha)} = M,$$
$$\| v_i \| \leqslant L\tau^2 iM \leqslant LM(\beta - \alpha)\tau$$

hold. From the last inequality it follows that when $\tau \to 0$,

$$\| v_i \| \to 0.$$

Thus, the proof of the convergence of the difference solution x_i to an accurate solution $x(i\tau)$ is completed.

Upon substituting now, on the basis of Corollary 1.2.1, the matrix A^+ (1.11.5) by its representation $A^+ = (A^*A)^{-1}A^*$, we obtain the difference scheme

$$A_i^*A_i \frac{x_i - x_{i-1}}{\tau} = A_i^*B_i x_i + A_i^* f_i, \quad i = 1, \dots, N; \qquad N\tau = \beta - \alpha, \quad (2.1.9)$$

$$x_0 = a,$$

suitable for the calculations. The recurrent formula, which must be used in this case, has the form

$$x_i = (A_i^*A_i - \tau A_i^*B_i)^{-1}(A_i^*A_i x_{i-1} + \tau A_i^* f_i), \quad i = 1, \dots, N;$$

$$x_0 = a.$$

Remark 2.1.1   Of course, scheme (2.1.9) could be obtained by straightforwardly multiplying equation (2.1.1) by the matrix A^*, but, as will be apparant later, our derivation is also suitable for the general case.

2.2.   The system with a regular pair of matrices

The pair of matrices $(A(t), B(t))$ is said to be regular on the segment $\alpha \leqslant t \leqslant \beta$, if there exists a number c at which, for all $t \in [\alpha, \beta]$, the matrix $B(t) - cA(t)$ is reversible (cf. [1, p. 332]).
    Let us consider the system

$$A(t)x'(t) = B(t)x(t) + f(t), \qquad (2.2.1)$$

in which the pair of matrices $(A(t), B(t))$ is regular.
    By substituting $(x(t) = e^{ct}y(t))$, the system (2.2.1) is brought to the form

$$(B - cA)^{-1}Ay' = y + e^{-ct}(B - cA)^{-1}f.$$

For this reason, when considering the Cauchy problems with a regular pair of matrices, one can limit oneself to examining problems of the form

$$A(t)x'(t) = x(t) + f(t), \quad \alpha \leqslant t \leqslant \beta, \qquad (2.2.2)$$

$$x(\alpha) = a. \qquad (2.2.3)$$

Lemma 2.2.1   For any solution x of the problem (2.2.2), (2.2.3), there will be a vector y such that the pair (x, y) satisfies the system

$$A^{\mathscr{D}}A^2x' = x - y + f, \qquad (2.2.4)$$

$$(E - A^{\mathscr{D}}A)Ay' = y - (E - A^{\mathscr{D}}A)A(A^{k-1})'(A^{\mathscr{D}})^{k-1}(x + f)$$
$$+ (E - A^{\mathscr{D}}A)Af', \qquad (2.2.5)$$

$$x(\alpha) = a,$$

$$y(\alpha) = [E - A^{\mathscr{D}}(\alpha)A(\alpha)][a + f(\alpha)], \qquad (2.2.6)$$

where k is an integer number not smaller than the index of the matrix $A(t)$ which is assumed to be independent of $t \in [\alpha, \beta]$.

Conversely, if the pair of vectors (x, y) satisfies the system (2.2.4)–(2.2.6), then the vector x is the solution of the problem (2.2.2), (2.2.3).

Proof Let x be the solution of problem (2.2.2), (2.2.3). Then, by multiplying the equality (2.2.2) by the matrix $A^{\mathscr{D}}A$, we obtain the equality (2.2.4), in which

$$y = (E - A^{\mathscr{D}}A)(x + f), \tag{2.2.7}$$

and, since the substitution of the vector (2.2.7) into equation (2.2.5), in view of the equalities (1.8.1), (1.8.13), and (2.2.2) leads to an identity, the first assertion of Lemma 2.2.1 is proved. In order to prove the inverse assertion, we multiply the equalities (2.2.4) and (2.2.5) by the matrix $A^{\mathscr{D}}A$. As a result, we have

$$A^{\mathscr{D}}A^2 x' = A^{\mathscr{D}}Ax - A^{\mathscr{D}}Ay + A^{\mathscr{D}}Af, \qquad A^{\mathscr{D}}Ay = 0,$$

from which it follows that

$$A^{\mathscr{D}}A^2 x' = A^{\mathscr{D}}Ax + A^{\mathscr{D}}Af. \tag{2.2.8}$$

By multiplying further the equality (2.2.4) by the matrix $E - A^{\mathscr{D}}A$, we obtain the expression (2.2.7) for y. Substitution of (2.2.7) into (2.2.5), in view of (1.8.1) and (1.8.13), leads to the equality

$$(E - A^{\mathscr{D}}A)Ax' = (E - A^{\mathscr{D}}A)x + (E - A^{\mathscr{D}}A)f, \tag{2.2.9}$$

and, since the sum of the equalities (2.2.8) and (2.2.9) is the equality (2.2.2), the second assertion of Lemma 2.2.1 is also proved (the reasoning regarding conditions (2.2.3) and (2.2.6) is obvious).

Systems of the form (2.2.2), in which the inequality ind $A(t) \leqslant k = \text{const}$ is satisfied for the index of the matrix $A(t)$, will be called the systems of index k, so that when $k_1 \leqslant k_2$ the system of index k_1 is simultaneously the system of index k_2 as well.

2.3. Systems of index 1

The matrix A in the system of index 1 satisfies the equality $A^{\mathscr{D}}A^2 = A$, as a consequence of which the equivalent problem (2.2.4)–(2.2.6) from the formulation of Lemma 2.2.1 is transformed into the initial problem

$$Ax' = x + f, \tag{2.3.1}$$

$$x(\alpha) = a, \tag{2.3.2}$$

with the condition of matching the initial data with the right-hand side

$$[E - A^{\mathscr{D}}(\alpha)A(\alpha)][a + f(\alpha)] = 0. \tag{2.3.3}$$

There have recently appeared in the literature quite a large number of interesting papers (see, for example, [17–19]) devoted to numerical (in particular, difference) methods of solving the problem (2.3.1), (2.3.2). The results of those publications induce one to think that, apparently, all known implicit methods of the approximate solution of a singular problem of the form (2.3.1), (2.3.2) are also applicable in the case when the system (2.3.1), although being singular, has an index equal to unity.

The simplest difference scheme for solving the problem (2.3.1), (2.3.2) (in the case when the index of the system is equal to unity) is the implicit Euler scheme

$$x_i = (A_i - \tau E)^{-1} A_i x_{i-1} + \tau (A_i - \tau E)^{-1} f_i,$$
$$x_0 = a, \quad i = 1, \ldots, N; \qquad N\tau = \beta - \alpha, \tag{2.3.4}$$

and, according to equality (2.3.3), the initial data $x_0 = a$ must satisfy the condition

$$(E - A_0^{\mathscr{D}} A_0) a = - (E - A_0^{\mathscr{D}} A_0) f_0.$$

In order to prove the convergence of the difference solution to an exact one, we introduce (as was done in [1]) into our consideration the deviation $v_i = x_i - x(i\tau)$ and substitute $x_i = v_i + x(i\tau)$ into the difference equation (2.3.4). In this case we obtain:

$$v_i = (A_i - \tau E)^{-2} A_i v_{i-1} + (A_i - \tau E)^{-1} [-A_i(x(i\tau) - x(i-1)\tau) + \tau x(i\tau) + \tau f(i\tau)],$$
$$v_0 = 0, \quad i = 1, \ldots, N; \qquad N\tau = \beta - \alpha. \tag{2.3.5}$$

Furthermore, we make use of formula (2.1.8) and bring equation (2.3.5) into the form

$$v_i = (A_i - \tau E)^{-1} A_i [v_{i-1} + (\tau^2/2) x''(i\theta\tau)]. \tag{2.3.6}$$

When $k = 1$, by applying the resolvents presented

$$(A - \tau E)^{-1} = - \sum_{i=1}^{k} \tau^{-i} A^{i-1} (E - A^{\mathscr{D}} A) + (E - \tau A^{\mathscr{D}})^{-1} A^{\mathscr{D}} \tag{2.3.7}$$

(see (1.8.16)), we obtain $(A - \tau E)^{-1} A = (E - \tau A^{\mathscr{D}})^{-1} A^{\mathscr{D}} A$, and, therefore, the equality (2.3.6) can be written as

$$v_i = (E - \tau A_i^{\mathscr{D}})^{-1} A_i^{\mathscr{D}} A_i [v_{i-1} + (\tau^2/2) x''(i\theta\tau)]. \tag{2.3.8}$$

By multiplying equality (2.3.8) by the matrix $E - A_i^{\mathscr{D}} A_i$, we obtain $(E - A_i^{\mathscr{D}} A_i) v_i = 0$ $(i = 1, \ldots, N)$ and, consequently, for the vector $(E - A_i^{\mathscr{D}} A_i) x_i$, the Euler scheme (2.3.4) yields an accurate value. In this case, since $(E - A_{i-1}^{\mathscr{D}} A_{i-1}) v_{i-1} = 0$, we can write

$$A_i^{\mathscr{D}} A_i v_{i-1} = (E + A_i^{\mathscr{D}} A_i - A_{i-1}^{\mathscr{D}} A_{i-1}) v_{i-1},$$

and, hence, equation (2.3.8) is brought to the form

$$v_i = C_i v_{i-1} + \tau^2 \varphi_i, \qquad v_0 = 0,$$

where

$$C_i = (E - \tau A_i^{\mathscr{D}})^{-1}(E + A_i^{\mathscr{D}} A_i - A_{i-1}^{\mathscr{D}} A_{i-1}),$$
$$\varphi_i = (\tfrac{1}{2})(E - \tau A_i^{\mathscr{D}})^{-1} A_i^{\mathscr{D}} A_i x''(i\theta\tau).$$

Now, as in [1], one can conclude that, as a consequence of the sufficient smoothness of the matrices $A^{\mathscr{D}}$ and $(A^{\mathscr{D}} A)'$ and of the vector $x(t)$ there exist constant K and L independent of $i = 1, \ldots, N$ and of τ from some interval $(0, \tau_0)$ such that

$$\|C_i\| \leqslant 1 + K\tau, \qquad \|\varphi_i\| \leqslant L,$$

and, therefore, $\|v_i\| \to 0$ when $\tau \to 0$.

Thus, the difference scheme (2.3.4) is convergent.

2.4.   Systems having the property Ω

If the matrix $(E - A^{\mathscr{D}} A)A$ is constant, the system (2.3.1) will be called a system having the property Ω.

It is obvious that in the preceding section we have considered systems of index 1 which are involved in the class of systems having the property Ω, as well as all systems (2.3.1), in which the matrix A is constant. The following lemma, which is a corollary of Lemma 2.2.1, is valid.

Lemma 2.4.1   If the system (2.3.1) has the property Ω and ind $A(t)\, k \leqslant$ constant, then the problem

$$Ax' = x + f, \tag{2.4.1}$$

$$x(\alpha) = a, \tag{2.4.2}$$

is equivalent to the problem

$$A^{\mathscr{D}} A^2 x' = x + (E - A^{\mathscr{D}} A)z + A^{\mathscr{D}} Af, \qquad x(\alpha) = a, \tag{2.4.3}$$

where $z = \sum_{s=0}^{k-1} A^s f^{(s)}$, and the condition for matching the initial data with the right-hand side is provided by the equality

$$(E - A^{\mathscr{D}} A)|_{t=\alpha}\, a = -(E - A^{\mathscr{D}} A)z|_{t=\alpha}. \tag{2.4.4}$$

Proof   Since the matrix $(E - A^{\mathscr{D}} A)A$ is constant and ind $A \leqslant k$, then

$$(E - A^{\mathscr{D}} A)A(A^{k-1})' = [(E - A^{\mathscr{D}} A)A^k]' = 0$$

and the middle summand on the right-hand side of equation (2.2.5) disappears. In this case, as can easily be verified, the sequence

$$y_v = (E - A^{\mathscr{D}} A)Ay'_{v-1} - (E - A^{\mathscr{D}} A)Af', \tag{2.4.5}$$

$$y_0 = 0, \quad v = 1, 2, \ldots,$$

starting with $v = k - 1$, as a consequence of the nilpotency of the matrix $(E - A^{\mathscr{D}}A)A$, stabilizes and leads to the solution of equation (2.2.5):

$$y_{k-1} = -(E - A^{\mathscr{D}}A) \sum_{s=1}^{k-2} A^s f^{(s)}$$

(further information regarding this can be obtained from [2] and [20]).

On substituting this solution into (2.2.4) and (2.2.6) we obtain (2.4.3) and (2.4.4). The lemma is thereby proved.

Since ind $A^{\mathscr{D}}A^2 \leqslant 1$, to solve the problem (2.4.3) one can apply the implicit Euler scheme

$$x_i = (A_i^{\mathscr{D}}A_i^2 - \tau E)^{-1}[A_i^{\mathscr{D}}A_i^2 x_{i-1} + \tau(E - A_i^{\mathscr{D}}A_i)z_i + \tau A_i^{\mathscr{D}}A_i f_i],$$

$$x_0 = a, \quad z_i = z(i\tau),$$

the convergence of which was proved in the preceding section and which by applying the equality (2.3.7), in view of the fact that now ind $A^{\mathscr{D}}A^2 \leqslant 1$ and $(A^{\mathscr{D}}A^2)^{\mathscr{D}} = A^{\mathscr{D}}$, can be written as

$$x_i = (E - \tau A_i^{\mathscr{D}})^{-1} A_i^{\mathscr{D}} A_i x_{i-1} - (E - A_i^{\mathscr{D}}A_i)z_i + \tau(E - \tau A_i^{\mathscr{D}})^{-1} A_i^{\mathscr{D}} f_i, \tag{2.4.6}$$

$$x_0 = a, \quad z_i = z(i\tau).$$

However, to perform the calculation according to scheme (2.4.6), one has to know the inverse Drasin matrix $A^{\mathscr{D}}$, which is not always determined in a straightforward way; therefore it is necessary to use the approximations

$$(A^{k+1} + \tau^2 E)^{-1} A^k \sim A^{\mathscr{D}}, \tag{2.4.7}$$

$$(A^{k+1} + \tau^2 E)^{-1} A^{k+1} \sim A^{\mathscr{D}}A, \tag{2.4.8}$$

$$\tau^2 (A^{k+1} + \tau^2 E)^{-1} \sim E - A^{\mathscr{D}}A, \tag{2.4.9}$$

following from Theorem 1.8.6. By applying these approximations (in view of (2.4.6)), one is led to the scheme

$$(A_i^{k+1} - \tau A_i^k + \tau^2 E)x_i = A_i^{k+1} x_{i-1} - \tau^2 z_i + \tau A_i^k f_i, \tag{2.4.10}$$

$$x_0 = a, \quad z_i = z(i\tau),$$

which is already suitable for our calculations.

In order to prove the convergence of the difference solution x_i to an accurate $x(i\tau)$, we estimate, as above, the deviation $v_i = x_i - x(i\tau)$, which in this case satisfies the equation

$$(A_i^{k+1} - \tau A_i^k + \tau^2 E)v_i = A_i^{k+1} v_{i-1} - (A_i^{k+1} - \tau A_i^k + \tau^2 E)x(i\tau)$$
$$+ A_i^{k+1} x((i-1)\tau) - \tau^2 z_i + \tau A_i^k f_i, \quad v_0 = 0. \tag{2.4.11}$$

By multiplying equation (2.4.11) by the matrix $E - A_i^{\mathscr{D}}A_i$, we obtain:

$$\tau^2 [E - A_i^{\mathscr{D}}A_i]v_i = -\tau^2 (E - A_i^{\mathscr{D}}A_i)[x(i\tau) + z_i],$$

and, since the right-hand side of this equality, by virtue of equation (2.4.3), is zero, then $(E - A_i^{\mathscr{D}} A_i)v_i = 0$. Thus, the difference scheme (2.4.10) for the vector $(E - A_i^{\mathscr{D}} A_i)x_i$ yields an accurate value.

In order to prove the convergence of $v_i \to 0$ when $\tau \to 0$, we shall make use of the fact that

$$(A^{k+1})^{\mathscr{D}} A^{k+1} = A^{\mathscr{D}} A, \qquad (A^{k+1})^{\mathscr{D}} A^{\mathscr{D}} A = (A^{k+1})^{\mathscr{D}}, \qquad (2.4.12)$$

$$(A^{k+1})^{\mathscr{D}} A^k = A^{\mathscr{D}} = A^{\mathscr{D}} A^{\mathscr{D}} A.$$

By multiplying the equality (2.4.11) by the matrix $(A^{k+1})^{\mathscr{D}}$, in view of the equalities (2.4.12) and $v_i = A_i^{\mathscr{D}} A_i v_i, v_{i-1} = A_{i-1}^{\mathscr{D}} A_{i-1} v_{i-1}$ for the vector v_i, we obtain:

$$[E - \tau A_i^{\mathscr{D}} + \tau^2 (A_i^{k+1})^{\mathscr{D}}]v_i$$
$$= (E + A_i^{\mathscr{D}} A_i - A_{i-1}^{\mathscr{D}} A_{i-1})v_{i-1} - [A_i^{\mathscr{D}} A_i - \tau A_i^{\mathscr{D}} + \tau^2 (A_i^{k+1})^{\mathscr{D}}]x(i\tau)$$
$$+ A_i^{\mathscr{D}} A_i x((i-1)\tau) - \tau^2 (A_i^{k+1})^{\mathscr{D}} z_i + \tau A_i^{\mathscr{D}} f_i.$$

If now (2.1.8) is substituted into the obtained equality, then, as in [3], we arrive at the equality

$$[E - \tau A_i^{\mathscr{D}} + \tau^2 (A_i^{k+1})^{\mathscr{D}}]v_i = (E + A_i^{\mathscr{D}} A_i - A_{i-1}^{\mathscr{D}} A_{i-1})v_{i-1}$$
$$+ \tau^2 [\tfrac{1}{2} A_i^{\mathscr{D}} A_i x''((i-1)\theta\tau) - (A_i^{k+1})^{\mathscr{D}} (x(i\tau) + z_i)],$$

from which it easily follows that $v_i \to 0$ when $\tau \to 0$. The proof of the convergence for the scheme (2.4.10) is completed.

2.5. Systems of index 2. Particular cases

A feature of systems of index 2 is the fact that equation (2.2.5) is written as an algebraic system. For example, the following lemma, which is the corollary from Lemma 2.2.1, is valid.

Lemma 2.5.1   Let ind $A \leqslant 2$ and

$$(E - A^{\mathscr{D}} A)AA'A^{\mathscr{D}} = 0. \qquad (2.5.1)$$

Then, for any solution x of the problem

$$Ax' = x + f, \qquad (2.5.2)$$

$$x(\alpha) = a, \qquad (2.5.3)$$

there will be a vector y such that the pair (x, y) satisfies the system

$$A^{\mathscr{D}} A^2 x' = x - y + f, \qquad (2.5.4)$$

$$Ay = 0, \qquad (2.5.5)$$

$$[E + (E - A^{\mathscr{D}}A)A']y = -(E - A^{\mathscr{D}}A)Af', \tag{2.5.6}$$

$$x(\alpha) = a, \qquad y(\alpha) = [E - A^{\mathscr{D}}(\alpha)A(\alpha)][a + f(\alpha)]. \tag{2.5.7}$$

Conversely, if the pair of vectors (x, y) satisfies the system (2.5.4)–(2.5.7), then the vector x is the solution of the problem (2.5.2), (2.5.3).

Proof We multiply equation (2.2.5) by the matrix A. As a result (as a consequence of the inequality $\operatorname{ind} A \leqslant 2$) we obtain $Ay = 0$, i.e. (2.5.5). Furthermore, if we use the equality $Ay' = -A'y$, which is a consequence of the equality $Ay = A'y$, as well as the condition (2.5.1), from equation (2.2.5) it is easy to switch over to equation (2.5.6). It is also easy to show that the system (2.5.4)–(2.5.7) gives rise to the system (2.2.4)–(2.2.6). Thus, Lemma 2.5.1 is valid.

Corollary 2.5.1 Let $\operatorname{ind} A \leqslant 2$,

$$(E - A^{\mathscr{D}}A)AA'A^{\mathscr{D}} = 0, \tag{2.5.8}$$

and the matrix

$$A^*A + [E + (E - A^{\mathscr{D}}A)A']^*[E + (E - A^{\mathscr{D}}A)A'] \tag{2.5.9}$$

be nonsingular (for all $t \in [\alpha, \beta]$). Then the solution of the problem (2.5.2), (2.5.3) exists if and only if the equalities

$$AC(E - A^{\mathscr{D}}A)Af' = 0,$$
$$[E + (E - A^{\mathscr{D}}A)A']C(E - A^{\mathscr{D}}A)Af' = (E - A^{\mathscr{D}}A)Af', \tag{2.5.10}$$

are satisfied as well as the condition for matching with the right-hand side of the initial data.

$$[E - A^{\mathscr{D}}(\alpha)A(\alpha)][a + f(\alpha)] + C(\alpha)[E - A^{\mathscr{D}}(\alpha)A(\alpha)]A(\alpha)f'(\alpha) = 0, \tag{2.5.11}$$

where

$$C = \{A^*A + [E + (E - A^{\mathscr{D}}A)A']^*[E + (E - A^{\mathscr{D}}A)A']\}^{-1}[E + (E - A^{\mathscr{D}}A)A']^*. \tag{2.5.12}$$

In this case the problem (2.5.2), (2.5.3) is equivalent to the problem

$$A^{\mathscr{D}}A^2x' = x - y + f, \tag{2.5.13}$$

$$y = -C(E - A^{\mathscr{D}}A)Af', \tag{2.5.14}$$

$$x(\alpha) = a, \tag{2.5.15}$$

and has a unique solution.

The proof is based on the following well-known result: if the matrix B in the algebraic system $By = \varphi$ is such that the matrix B^*B is nonsingular, then the system is compatible if and only if $B(B^*B)^{-1}B^*\varphi = \varphi$, and the solution of the system is unique and is expressed by the formula $y = (B^*B)^{-1}B^*\varphi$ (see also

Corollaries 1.2.1 and 1.3.1 and Lemma 1.4.13). By applying this result to the system (2.5.5)–(2.5.7), it is easy to obtain the compatibility conditions (2.5.9), (2.5.10), the matching condition (2.5.11), and formula (2.5.14).

The condition (2.5.8), under which the results formulated (Lemma 2.5.1 and Corollary 2.5.1) are proved, is satisfied, in particular, in the case of constancy of either the matrix $(E - A^{\mathscr{D}}A)A$ or the matrix $A^{\mathscr{D}}A$. The former case has already been considered from a different point of view in the preceding two sections. As far as the latter case is concerned, it too deserves special attention. The point here is that if the matrix $A^{\mathscr{D}}A$ is constant, then as a consequence of the fact that $A^{\mathscr{D}}AA' = A'A^{\mathscr{D}}A$ and $Ay = 0$, the system (2.5.5), (2.5.6) simplifies to become

$$Ay = 0, \qquad (E + A')y = -(E - A^{\mathscr{D}}A)Af'.$$

In this case Corollary 2.5.1 receives a simpler formulation:

Corollary 2.5.2 Let ind $A \leqslant 2$, the matrix $A^{\mathscr{D}}A$ be constant, and the matrix

$$A^*A + (E + A')^*(E + A') \tag{2.5.16}$$

be nonsingular. Then the solution of the problem (2.5.2), (2.5.3) exists if and only if the equalities

$$AC(E - A^{\mathscr{D}}A)Af' = 0,$$
$$(E + A')C(E - A^{\mathscr{D}}A)Af = (E - A^{\mathscr{D}}A)f', \tag{2.5.17}$$

and the matching condition (2.5.11) are satisfied, in which

$$C = [A^*A + (E + A')^*(E + A')]^{-1}(E + A')^*. \tag{2.5.18}$$

In this case the solution of the problem (2.5.2), (2.5.3) is unique and equivalency of the problems (2.5.2), (2.5.3) and (2.5.13)–(2.5.15) takes place.

On the basis of Corollaries 2.5.1 and 2.5.2 it is possible to construct stable and difference schemes in the following way: we write, first, the implicit difference scheme

$$(A_i^{\mathscr{D}}A_i^2 - \tau E)x_i = A_i^{\mathscr{D}}A_i^2 x_{i-1} - \tau y_i + \tau f_i, \tag{2.5.19}$$

which approximates equation (2.5.13), and then multiply equation (2.5.19) by the matrix

$$(E - \tau A_i^{\mathscr{D}})(A_i^{\mathscr{D}}A_i^2 - \tau E)^{-1} = -\frac{1}{\tau}(E - A_i^{\mathscr{D}}A_i) + A_i^{\mathscr{D}}$$

(see (2.3.7)) and, as a result, we obtain:

$$(E - \tau A_i^{\mathscr{D}})x_i = A_i^{\mathscr{D}}A_i x_{i-1} + y_i - (E - A_i^{\mathscr{D}}A_i)f_i + \tau A_i^{\mathscr{D}}f_i. \tag{2.5.20}$$

Finally, we replace the matrix $A_i^{\mathscr{D}}$ in the equality (2.5.20) (and also in the final stage in the equality (2.5.14)) with its approximation

$$(A_i^3 + \tau^2 E)^{-1}A_i^2.$$

The resulting difference scheme has the form

$$(A_i^3 - \tau A_i^2 + \tau^2 E)x_i = A_i^3 x_{i-1} + \tau^2 y_i - \tau^2 f_i + \tau A_i^2 f_i, \tag{2.5.21}$$

$$y_i = - \tau^2 \tilde{C}_i (A_i^3 + \tau^2 E)^{-1} A_i f_i', \tag{2.5.22}$$

$$x_0 = a, \tag{2.5.23}$$

where

$$\tilde{C}_i = \{A_i^* A_i + [E + \tau^2 (A_i^3 + \tau^2 E)^{-1} A_i']^* [E + \tau^2 (A_i^3 + \tau^2 E)^{-1} A_i']\}^{-1}$$
$$\times [E + \tau^2 (A_i^3 + \tau^2 E)^{-1} A_i']^* \tag{2.5.24}$$

or

$$\tilde{C}_i = C_i = [A_i^* A_i + (E + A_i')^* (E + A_i')]^{-1} (E + A_i')^* \tag{2.5.25}$$

if the matrix $A^{\mathscr{D}} A$ is constant.

In order to prove the convergence of the difference solution to an accurate one, we introduce into our treatment the deviation $v_i = x_i - x(i\tau)$. The deviation v_i, in view of formula (2.1.8) and the equality (2.5.4), satisfies the equation

$$(A_i^3 - \tau A_i^2 + \tau^2 E)v_i = A_i^3 v_{i-1} - \tau^2 A_i^{\mathscr{D}} A_i^2 x'(i\tau)$$
$$+ \tfrac{1}{2}\tau^2 A_i^3 x''((i-1)\theta\tau) + \tau^2 [y_i - y(i\tau)]. \tag{2.5.26}$$

By multiplying this equation by the matrix $E - A_i^{\mathscr{D}} A_i$ we obtain:

$$(E - A_i^{\mathscr{D}} A_i)v_i = (E - A_i^{\mathscr{D}} A_i)[y_i - y(i\tau)]. \tag{2.5.27}$$

Now, if the equality

$$(A^3 + \tau^2 E)^{-1} = \frac{1}{\tau^2}(E - A^{\mathscr{D}} A) + [E + \tau^2 (A^{\mathscr{D}})^3]^{-1}(A^{\mathscr{D}})^3 \tag{2.5.28}$$

is taken into account as well as expressions (2.5.22) and (2.5.14), after making long calculations (it is pointless to give them here) for the difference $y_i - y(i\tau)$ and the vector (2.5.27), one can obtain the following estimates:

$$\| y_i - y(i\tau) \| \leqslant M\tau^2, \tag{2.5.29}$$

$$\| (E - A_i^{\mathscr{D}} A_i)v_i \| \leqslant K\tau^2. \tag{2.5.30}$$

Furthermore, in order to estimate the vector $\eta_i = A_i^{\mathscr{D}} A_i v_i$, we multiply the equality (2.5.26) by the matrix $(A_i^3)^{\mathscr{D}}$. As a result, taking into account the obvious equality,

$$A_i^{\mathscr{D}} A_i = (E + A_i^{\mathscr{D}} A_i - A_{i-1}^{\mathscr{D}} A_{i-1})A_{i-1}^{\mathscr{D}} A_{i-1} + A_i^{\mathscr{D}} A_i (E - A_{i-1}^{\mathscr{D}} A_{i-1}),$$

as well as the equality (2.5.27), for η_i we obtain the equation

$$[E - \tau A_i^{\mathscr{D}} + \tau^2 (A_i^3)^{\mathscr{D}}]\eta_i = (E + A_i^{\mathscr{D}} A_i - A_{i-1}^{\mathscr{D}} A_{i-1})\eta_{i-1}$$
$$- \tau^2 (A_i^{\mathscr{D}})^2 x'(i\tau) + \tfrac{1}{2}\tau^2 A_i^{\mathscr{D}} A_i x''((i-1)\theta\tau)$$
$$+ [A_i^{\mathscr{D}} A_i (E - A_{i-1}^{\mathscr{D}} A_{i-1}) + \tau^2 (A_i^{\mathscr{D}})^3][y_i - y(i\tau)],$$
$$\eta_0 = 0.$$

Using this, as in the previous section, and in view of (2.5.29), it is easy to prove that the vector $\eta_i = A_i^{\mathscr{D}} A_i v_i$ tends to zero when τ tends to zero. But then, as a consequence of (2.5.30) and of the inequality

$$\| v_i \| \leqslant \| (E - A_i^{\mathscr{D}} A_i) v_i \| + \| A_i^{\mathscr{D}} A_i v_i \|,$$

the same is valid for the vector v_i: $v_i \to 0$ when $\tau \to 0$. The convergence of the difference scheme (2.5.21)–(2.5.23) is thereby substantiated.

Remark 2.5.1   One should not disregard the case where the matrices (2.5.9) and (2.5.16) are nonsingular as a consequence of the nonsingularity of the matrices $E + (E - A^{\mathscr{D}} A) A'$ and $E + A'$. In this case the vector y is uniquely determined from (2.5.6), and the formulae (2.5.12), (2.5.16), (2.5.24), and (2.5.25), for calculating the matrices C and $\tilde{C}_i$, assume a simpler form:

$$C = [E + (E - A^{\mathscr{D}} A) A']^{-1},$$
$$C = (E + A')^{-1},$$
$$\tilde{C}_i = [E + \tau^2 (A_i^3 + \tau^2 E)^{-1} A_i']^{-1},$$
$$\tilde{C}_i = (E + A_i')^{-1}.$$

In this case the existence conditions for the solution associated with the requirement for the validity of the equality $Ay = 0$ are satisfied for any vector f' (the verification is left to the reader). When verifying, it is useful to take into consideration that, if the index of the matrix A does not exceed two, then the equality

$$(E - A^{\mathscr{D}} A) A A'(E - A^{\mathscr{D}} A) = -(E - A^{\mathscr{D}} A) A' A (E - A^{\mathscr{D}} A)$$

holds, which follows from the chain

$$0 = [(E - A^{\mathscr{D}} A) A^2]'(E - A^{\mathscr{D}} A) = -(A^{\mathscr{D}} A)' A^2 (E - A^{\mathscr{D}} A)$$
$$+ (E - A^{\mathscr{D}} A)(A'A + AA')(E - A^{\mathscr{D}} A) = (E - A^{\mathscr{D}} A)(A'A + AA')(E - A^{\mathscr{D}} A).$$

2.6.   Systems of index 2. The general case

When examining the general case, a different formulation of Lemma 2.2.1 turns out more preferable.

Lemma 2.6.1   Any solution of the problem

$$Ax' = x + f, \tag{2.6.1}$$

$$x(\alpha) = a, \tag{2.6.2}$$

is represented as

$$x = y + z, \tag{2.6.3}$$

where the vectors y and z satisfy the system

$$A^{\mathscr{D}}A^2z' = z + [A^{\mathscr{D}}AA' + (A^{\mathscr{D}})^{k-1}(A^{k-1})'A]y + f, \tag{2.6.4}$$

$$(E - A^{\mathscr{D}}A)Ay' = y - (E - A^{\mathscr{D}}A)A(A^{k-1})'(A^{\mathscr{D}})^{k-1}(z + f) + (E - A^{\mathscr{D}}A)Af', \tag{2.6.5}$$

$$y(\alpha) = [E - A^{\mathscr{D}}(\alpha)A(\alpha)][a + f(\alpha)],$$
$$z(\alpha) = A^{\mathscr{D}}(\alpha)A(\alpha)a - [E - A^{\mathscr{D}}(\alpha)A(\alpha)]f(\alpha). \tag{2.6.6}$$

Conversely, the vector x that is the sum (2.6.3), in which all vectors y and z satisfy the system (2.6.4)–(2.6.6), is the solution of the problem (2.6.1), (2.6.2).

Proof This is reduced to substituting into the system (2.2.4)–(2.2.6), in view of the equalities (1.8.1), (1.8.13) and $A^{\mathscr{D}}Ay = 0$.

The general case of the systems of index 2 will be examined here on the basis of the formulated Lemma 2.6.1, by nevertheless making the assumption that the matrix

$$A^*A + [E + (E - A^{\mathscr{D}}A)A']^*[E + (E - A^{\mathscr{D}}A)A']$$

will be considered nonsingular. Under such an assumption using the same methods as used when proving Corollary 2.5.1, from Lemma 2.5.1 one can obtain the following result.

Corollary 2.6.1 (from Lemma 2.6.1) Let ind $A \leqslant 2$. Then, any solution of the problem (2.6.1), (2.6.2) is represented as

$$x = y + z, \tag{2.6.7}$$

where the vectors y and z satisfy the system

$$A^{\mathscr{D}}A^2z' = [E + A^{\mathscr{D}}AA'C(E - A^{\mathscr{D}}A)AA'A^{\mathscr{D}}]z$$
$$+ A^{\mathscr{D}}AA'C(E - A^{\mathscr{D}}A)A(A'A^{\mathscr{D}}f - f') + f, \tag{2.6.8}$$

$$AC(E - A^{\mathscr{D}}A)A[A'A^{\mathscr{D}}(z + f) - f'] = 0, \tag{2.6.9}$$

$$[E + (E - A^{\mathscr{D}}A)A']C(E - A^{\mathscr{D}}A)A[A'A^{\mathscr{D}}(z + f) - f']$$
$$= (E - A^{\mathscr{D}}A)A[A'A^{\mathscr{D}}(z + f) - f'], \tag{2.6.10}$$

$$y = C(E - A^{\mathscr{D}}A)A[A'A^{\mathscr{D}}(z + f) - f'], \tag{2.6.11}$$

where

$$C = \{A^*A + [E + (E - A^{\mathscr{D}}A)A']^*[E + (E - A^{\mathscr{D}}A)A']\}^{-1}$$
$$\times [E + (E - A^{\mathscr{D}}A)A']^* \tag{2.6.12}$$

(the initial data $y(\alpha)$ and $z(\alpha)$ are calculated by formulae (2.6.6)).

Conversely, the vector x calculated by formula (2.6.7), under the conditions (2.6.8)–(2.6.11) and (2.6.6), is the solution of the problem (2.6.1), (2.6.2).

A property of the system (2.6.8) is the fact that its pair of matrices has a special form

$$(A^{\mathscr{D}}A^2, E + A^{\mathscr{D}}AMA^{\mathscr{D}}), \tag{2.6.13}$$

from which follows the simplicity of the canonical structure of this pair. Indeed, let

$$A = N \begin{pmatrix} J_0 & 0 \\ 0 & J_1 \end{pmatrix} N^{-1}$$

be a Jordan representation of the matrix A (J_0 is the set of all nilpotent blocks) and

$$N^{-1}MN = \begin{pmatrix} M_1 & M_2 \\ M_3 & M_4 \end{pmatrix}.$$

Then, the pair of matrices (2.6.13) is brought to the following canonical form:

$$\left(\begin{pmatrix} 0 & 0 \\ 0 & J_1 \end{pmatrix}, \begin{pmatrix} E & 0 \\ 0 & E + M_4 J_1^{-1} \end{pmatrix} \right), \tag{2.6.14}$$

and, since (2.6.14) does not contain nilpotent blocks of order higher than 1, the initial pair of matrices (2.6.13) has a simple structure.

Systems, in which the structure of a pair of matrices is simple, via the substitution $z = e^{ct}z_1$, are brought to equivalent systems of the form $\tilde{A}z_1' = z_1 + \varphi$, the index of which equals unity and, therefore, as will be shown in the next chapter, allow for a simple analytic solution, unique in the case of the Cauchy problem. But then, by knowing the solution of the problem (2.6.8), (2.6.6), one can substitute this solution into equations (2.6.9) and (2.6.10) to obtain, in this way, the existence conditions for the solution of the initial problem (2.6.1), (2.6.2). We note, incidentally, that the result contained here on the surface, implies that if

$$AC(E - A^{\mathscr{D}}A)AA'A^{\mathscr{D}} = 0,$$
$$[E + (E - A^{\mathscr{D}}A)A']C(E - A^{\mathscr{D}}A)AA'A^{\mathscr{D}} = (E - A^{\mathscr{D}}A)AA'A^{\mathscr{D}},$$

then equations (2.6.9) and (2.6.10) immediately become the compatibility conditions.

Our assumption about the existence and uniqueness of the solution of the problems considered in this chapter on the basis of Corollary 2.6.1, allow us to conclude that the solution of the problem (2.6.1), (2.6.2) is reduced to solving the problem

$$A^{\mathscr{D}}A^2z' = [E + A^{\mathscr{D}}AA'C(E - A^{\mathscr{D}}A)AA'A^{\mathscr{D}}]z$$
$$+ A^{\mathscr{D}}AA'C(E - A^{\mathscr{D}}A)A(A'A^{\mathscr{D}}f - f') + f, \tag{2.6.15}$$
$$z(\alpha) = A^{\mathscr{D}}(\alpha)A(\alpha)a - [E - A^{\mathscr{D}}(\alpha)A(\alpha)]f(\alpha), \tag{2.6.16}$$

and the solution of the initial problem (2.6.1), (2.6.2) is obtained by the formula

$$x = [E + C(E - A_i^{\mathscr{D}}A)AA'A^{\mathscr{D}}]z + C(E - A^{\mathscr{D}}A)A(A'A^{\mathscr{D}}f - f') \quad (2.6.17)$$

(see (2.6.6)–(2.6.8) and (2.6.11)).

We start construction of the difference scheme suitable for solving the problem (2.6.15), (2.6.16) by first approximating the problem (2.6.15), (2.6.16) with the difference problem of the first order of accuracy

$$[A_i^{\mathscr{D}}A_i^2 - \tau E - \tau A_i^{\mathscr{D}}A_iA'_iC_i(E - A_i^{\mathscr{D}}A_i)A_iA'_iA_i^{\mathscr{D}}]z_i = A_i^{\mathscr{D}}A_i^2 z_{i-1}$$
$$+ \tau[A_i^{\mathscr{D}}A_iA'_iC_i(E - A_i^{\mathscr{D}}A_i)A_i(A'_iA_i^{\mathscr{D}}f_i - f'_i) + f_i], \quad (2.6.18)$$

$$z_0 = A_0^{\mathscr{D}}A_0 a - [E - A_0^{\mathscr{D}}A_0]f_0. \quad (2.6.19)$$

After that, making use of the equality

$$(E - \tau A_i^{\mathscr{D}})(A_i^{\mathscr{D}}A_i^2 - \tau E)^{-1} = -\frac{1}{\tau}(E - A_i^{\mathscr{D}}A_i) + A_i^{\mathscr{D}},$$

we bring equation (2.6.18) to the form

$$[E - \tau A_i^{\mathscr{D}} - \tau A_i^{\mathscr{D}}A'_iC_i(E - A_i^{\mathscr{D}}A_i)A_iA'_iA_i^{\mathscr{D}}]z_i$$
$$= A_i^{\mathscr{D}}A_i z_{i-1} - (E - A_i^{\mathscr{D}}A_i)f_i + \tau A_i^{\mathscr{D}}f_i$$
$$+ \tau A_i^{\mathscr{D}}A'_iC_i(E - A_i^{\mathscr{D}}A_i)A_i(A'_iA_i^{\mathscr{D}}f_i - f'_i).$$

Now if the approximations (2.4.7)–(2.4.9) (when $k = 2$) and (2.6.19) are taken into account, we then arrive at the difference scheme suitable for the calculations:

$$[A_i^3 - \tau A_i^2 + \tau^2 E - \tau^3 A_i^2 A'_i \tilde{C}_i T_i A_i A'_i T_i A_i^2]z_i$$
$$= A_i^3 z_{i-1} + \tau[A_i^2 - \tau E + \tau^2 A_i^2 A'_i \tilde{C}_i T_i A_i A'_i T_i A_i^2]f_i$$
$$- \tau^3 A_i^2 A'_i \tilde{C}_i T_i A_i f'_i, \quad (2.6.20)$$

$$z_0 = T_0(A_0^3 a - \tau^2 f_0), \quad i = 1,\ldots,N, \quad (2.6.21)$$

where

$$T_i = (A_i^3 + \tau^2 E)^{-1}, \quad (2.6.22)$$

$$\tilde{C}_i = [A_i^* A_i + (E + \tau^2 T_i A'_i)^*(E + \tau^2 T_i A'_i)]^{-1}(E + \tau^2 T_i A'_i)^*. \quad (2.6.23)$$

The approximate solution of the initial problem (2.6.1), (2.6.2) is calculated by the formula

$$x_i = [E + \tau^2 \tilde{C}_i T_i A_i A'_i T_i A_i^2]z_i + \tau^2 \tilde{C}_i T_i A_i A'_i T_i A_i^2 f_i - \tau^2 \tilde{C}_i T_i A_i f'_i. \quad (2.6.24)$$

In order to prove the convergence of z_i to $z(i\tau)$, we multiply the equalities (2.6.20) and (2.6.21) by the matrices $E - A_i^{\mathscr{D}}A_i$ and $E - A_0^{\mathscr{D}}A_0$, respectively. As a result, taking into account the equalities (2.5.28) and $(E - A^{\mathscr{D}}A)A^2 = 0$, for all $i = 0,\ldots,N$, we obtain:

$$(E - A_i^{\mathscr{D}}A_i)z_i = -(E - A_i^{\mathscr{D}}A_i)f_i.$$

But the same is also obtained when equation (2.6.8) and the condition (2.6.6) for $z(\alpha)$ is multiplied by the matrix $E - A^{\mathscr{D}}A$; therefore, for the deviation $v_i = z_i - z(i\tau)$ we have

$$(E - A_i^{\mathscr{D}}A_i)v_i = 0, \quad i = 0,\ldots,N, \tag{2.6.25}$$

which means that the vector $(E - A_i^{\mathscr{D}}A_i)z_i$ is calculated accurately according to the difference scheme (2.6.20), (2.6.21).

Now, if equation (2.6.20) is multiplied by the matrix $(A_i^{\mathscr{D}})^3$, and the equalities (2.6.12), (2.6.22), (2.6.23), (2.6.15), (2.6.16), (2.6.20), (2.6.21), and (2.6.25), and Theorem 1.8.6, are taken into account, then for the deviation v_i we obtain:

$$[E - \tau A_i^{\mathscr{D}} + \tau^2(A_i^{\mathscr{D}})^3 - \tau^3 A_i^{\mathscr{D}}A_i'C_iT_iA_iA_i'T_iA_i^2]v_i$$
$$= (E + A_i^{\mathscr{D}}A_i - A_{i-1}^{\mathscr{D}}A_{i-1})v_{i-1} + \varphi_i, \tag{2.6.26}$$

$$i = 1,\ldots,N, \quad v_0 = \varphi_0, \tag{2.6.27}$$

where $\|\varphi_i\| = O(\tau^2)$, $i = 0, 1,\ldots,N$ (we leave the calculations to the reader).

From equation (2.6.26) and the condition (2.6.27) it follows that $v_i \to 0$ when $\tau \to 0$. This result can be proved in the same way as was done in [1]. The only difference to be taken into account implies that, as a consequence of (2.6.27), $v_0 \ne 0$; however, this does not cause any complications.

Thus, it can be considered proved that $z_i = z(i\tau) + g_i$, where $\|g_i\| = O(\tau)$. But then $\|x_i - x(i\tau)\| = O(\tau)$ because formula (2.6.24), by virtue of Theorem 1.8.6, is an approximation of the second order of accuracy in τ. This completes the proof of the convergence of x_i to $x(i\tau)$ when $\tau \to 0$.

Remark 2.6.1   If the matrix

$$A^*A + [E + (E - A^{\mathscr{D}}A)A']^*[E + (E - A^{\mathscr{D}}A)A']$$

is nonsingular as a consequence of the nonsingularity of the matrix $E + (E - A^{\mathscr{D}}A)A'$, then as matrices C and $\tilde{C}_i$ (see (2.6.12), (2.6.22), and (2.6.23)) one can take, respectively, the matrices

$$C = [E + (E - A^{\mathscr{D}}A)A']^{-1},$$
$$\tilde{C}_i = [E + \tau^2(A_i^3 + \tau^2E)^{-1}A_i']^{-1}.$$

Equations (2.6.9) and (2.6.10) in this case are satisfied at any z, f, and f' (see, for an analogy, Remark 2.5.1).

2.7.   On the indices of matrices. Stable matrices

The results of the preceding sections lead to the need to establish simple signatures, in the presence of which the matrix index does not exceed some other number. In this regard, the following theorems hold true.

Theorem 2.7.1   The index of matrix A does not exceed an integer number $k \geqslant 0$

if and only if the equalities

$$A^k = A^k(A^{k+1})^- A^{k+1}, \qquad A^k = A^{k+1}(A^{k+1})^- A^k \qquad (2.7.1)$$

are satisfied.

Proof If ind $A \leqslant k$, then by virtue of the definition of the matrix index, the equality

$$\operatorname{rank} A^{k+1} = \operatorname{rank} A^k \qquad (2.7.2)$$

holds. But then, according to Lemmas 1.4.9 and 1.4.10 (see equalities (1.4.18) and (1.4.20)), equalities (2.7.1) also are valid. If, however, equalities (2.7.1) are satisfied, then, according to the same lemmas, equality (2.7.2) is valid.

Remark 2.7.1 It is easy to see that equalities (2.7.1) follow from one another; therefore, when applying them, it is sufficient to verify only one of them.

The algorithm for determining the matrix index is also contained in Theorem 1.9.1. The simplest corollary from it is thus.

Theorem 2.7.2 Let $A = B\Gamma$ be a skeleton expansion of the matrix A. Then, the nonsingularity of the matrix ΓB is a necessary and sufficient condition for the index of the matrix A not to exceed unity.

To formulate the following two theorems, which will indicate not only the index of the matrix but also its stability, we adopt the following agreement: the inequality $A \geqslant 0$ (or $A \leqslant 0$) means that the matrix A is positively (or negatively) semi-periodic, i.e. for it $(Ax, x) \geqslant 0$ (or $(Ax, x) \leqslant 0$) for all $x \in H$, where H is a unitary space of n-dimensional vectors with the scalar product (x, y) and norm $\|x\| = \sqrt{(x, x)}$. By a stable matrix we mean a matrix the nonzero eigen-numbers of which all lie in the left-hand semi-plane, and the index does not exceed unity.

Theorem 2.7.3 If the matrix A is such that $A^* + A \geqslant 0$ (or $A^* + A \leqslant 0$), then (1) the real parts of all eigen-numbers of the matrix A are non-negative (nonpositive); (2) $A^{\mathscr{D}} = A^+$; (3) the index of the matrix A does not exceed unity; and (4) if the rigorous inequality $A^* + A > 0$ (or $A^* + A < 0$) is satisfied, then the matrix A does not have purely imaginary and zero eigen-numbers.

Proof We consider only the case $A^* + A \geqslant 0$ (the case $A^* + A \leqslant 0$ can be considered in a similar way if it is taken into consideration that in this case $-(A^* + A) \geqslant 0$).

The validity of assertion (1) of the theorem is widely known. In order to prove assertions (2) and (3), we first remember that

$$2\operatorname{Re}(Ax, x) = ((A^* + A)x, x), \qquad 2\operatorname{Im}(Ax, x) = i((A^* - A)x, x), \qquad (2.7.3)$$

where the symbols Re and Im denote, respectively, the real and imaginary parts of the number.

Suppose now that $Ax = 0$. Then $(Ax, x) = 0$ and, as a consequence of (2.7.3), $(Sx, x) = 0$, where $S = A^* + A$. Furthermore, we note that since the matrix S is self-conjugate and, according to the condition of the theorem is positively semi-determinant, then, as is known, there exists a unique positively semi-determinant root from the matrix S. We denote it by $S^{1/2}$. As a result, we may write

$$0 = (Sx, x) = (S^{1/2}x, S^{1/2}x) = \| S^{1/2}x \|^2.$$

But then $S^{1/2}x = 0$ and, consequently, $Sx = (A^* + A)x = 0$, and since $Ax = 0$, then $A^*x = 0$. Thus, it is proved that $\ker A \subseteq \ker A^+$, i.e. $A^*(E - A^+A) = 0$ from which—in view of the self-conjugacy of the matrix A^+A, it follows that $(E - A^+A)A = 0$.

In much the same way, taking into account the equality $(A^*)^+ = (A^+)^*$, it is possible to show that $A(E - AA^+) = 0$.

Thus, if $A^* + A \geqslant 0$, then

$$A^+A^2 = A, \qquad A^2A^+ = A. \tag{2.7.4}$$

By multiplying the first of the equalities (2.7.4) by A^+ from the right and the second, by the matrix A^+, from the left, we obtain $AA^+ = A^+A$. But then, by multiplying the first of the equalities (2.7.4) by the matrix A^+ from the right and the left, we arrive at the equality

$$A^+ = A^+A^+AAA^+ = A^+AA^+AA^+ = A^+AA^+.$$

Thus, we obtain the system

$$AA^+ = A^+A, \qquad A^+AA^+ = A^+, \qquad A^+A^2 = A, \tag{2.7.5}$$

which is known to define the inverse Drasin matrix and, therefore, $A^+ = A^{\mathscr{D}}$. Thus, as a consequence of the last of the equalities (2.7.5), the index of the matrix A does not exceed unity. Assertions (2) and (3) are thereby proved.

Finally, we prove assertion (4). Suppose that $Ax = i\beta x$, where β is a real number. Then

$$((A^* + A)x, x) = (Ax, x) + (x, Ax) = i\beta \| x \|^2 - i\beta \| x \|^2 = 0,$$

which contradicts the inequality $A^* + A > 0$. Theorem 2.7.3 is proved completely.[†]

The proved theorem contains a sufficient signature of the stability of the matrix A: if $A^* + A \leqslant 0$, then the matrix A is stable. A necessary and sufficient signature of the stability of the matrix A can be obtained with the help of the

---

[†]Assertions (1), (3), and (4) can also be obtained from known results of the stability theory of linear ordinary differential equations [1].

following lemma, which is a generalization of the known Lyapunov result for the case of degenerate matrices.

Lemma 2.7.1 In order for all nonzero eigen-numbers of the matrix A to have a negative real part, it is necessary and sufficient that the equation

$$A^*Y + YA = -(A^{\mathscr{D}}A)^* A^{\mathscr{D}}A \tag{2.7.6}$$

has the solution Y, which is the Hermite matrix, positively defined on the transform of the matrix $A^{\mathscr{D}}A$.

Necessity Putting $f(z) = e^{zt}$ in (1.8.28), we obtain:

$$e^{At} = \sum_{j=1}^{m} \sum_{s=1}^{v_j} \frac{t^{s-1}}{(s-1)!} e^{\lambda_j t} A_j^{s-1}(E - A_j^{\mathscr{D}} A_j), \tag{2.7.7}$$

where $\lambda_j (j = 1, \ldots, m)$ are all nonzero eigen-numbers of the matrix A (if the matrix A is degenerate, then we put $\lambda_1 = 0$).

In order to exclude the zero eigen-number, we multiply the equality (2.7.7) by the matrix $A^{\mathscr{D}}A$. As a result (as a consequence of the equalities $A_0 = A, (A^{\mathscr{D}}A)^2 = A^{\mathscr{D}}A$), we obtain:

$$A^{\mathscr{D}}Ae^{At} = \sum_{j=2}^{m} \sum_{s=1}^{v_j} \frac{t^{s-1}}{(s-1)!} e^{\lambda_j t} A^{\mathscr{D}}A A_j^{s-1}(E - A_j^{\mathscr{D}} A_j). \tag{2.7.8}$$

If $\operatorname{Re} \lambda_j < 0$ when $j = 2, \ldots, m$ (and we do assume this) then obviously

$$A^{\mathscr{D}}Ae^{At} \to 0 \tag{2.7.9}$$

when $t \to \infty$, because when $t \to \infty$, all terms in (2.7.8) tend to zero. Moreover, when $\operatorname{Re} \lambda_j < 0$, the integrals

$$\int_0^\infty t^{s-1} e^{\lambda_j t} \, dt$$

are known to exist. But then also the integral

$$Y = \int_0^\infty e^{A^* t}(A^{\mathscr{D}}A)^* A^{\mathscr{D}}A e^{At} \, dt \tag{2.7.10}$$

exists. Taking this into account, we examine the matrix $Z(t)$ defined as the solution of the problem

$$Z' = A^*Z + ZA, \qquad Z(0) = -(A^{\mathscr{D}}A)^*(A^{\mathscr{D}}A), \tag{2.7.11}$$

and, consequently, having the form

$$Z(t) = -e^{A^* t}(A^{\mathscr{D}}A)^* A^{\mathscr{D}}A e^{At}.$$

If equation (2.7.11) is integrated from $t = \infty$ to $t = 0$, we obtain the equality (2.7.6), in which the matrix Y is determined by formula (2.7.10). Moreover, the matrix (2.7.10), obviously, is Hermitian and on the transform of the matrix $A^{\mathscr{D}}A$ is positively defined because, for any vector $x \neq 0$ having the form $x = A^{\mathscr{D}}Ay$, the inequality $(Yx, x) > 0$ holds true. Indeed, as a consequence of the validity of the equalities $(A^{\mathscr{D}}A)^2 = A^{\mathscr{D}}A$ and $A^{\mathscr{D}}Ae^{At} = e^{At}A^{\mathscr{D}}A$, we have

$$(Yx, x) = (YA^{\mathscr{D}}Ay, A^{\mathscr{D}}Ay) = \int_0^\infty ((A^{\mathscr{D}}A)^* e^{A^*t} e^{At} A^{\mathscr{D}}Ay, A^{\mathscr{D}}Ay) \, dt$$

$$= \int_0^\infty (e^{At} A^{\mathscr{D}}Ay, e^{At} A^{\mathscr{D}}Ay) \, dt = \int_0^\infty \| e^{At} x \|^2 \, dt > 0.$$

The necessity is proved.

Sufficiency Let equation (2.7.6) have the solution Y with the properties mentioned in the formulation of the theorem. Let us consider the behaviour of the Hermitian $u = (Yx, x)$ on the solutions of the system $x' = Ax$ with initial data of the form $x(0) = A^{\mathscr{D}}A\alpha$, where α is an arbitrary vector. Each such solution has the form $x(t) = e^{At} A^{\mathscr{D}}A\alpha$, and therefore the identity $(E - A^{\mathscr{D}}A)x(t) \equiv 0$ holds for it. But then we have

$$u' = (Yx', x) + (Yx, x') = (YAx, x) + (Yx, Ax) = ((YA + A^*Y)x, x)$$

$$= -((A^{\mathscr{D}}A)^* A^{\mathscr{D}}Ax, x) = -(A^{\mathscr{D}}Ax, A^{\mathscr{D}}Ax) = -(x, x) = -\|x\|^2. \quad (2.7.12)$$

Furthermore, we consider the function $\varphi(y) = (Yy, y)$, defined on a unit sphere K lying in the transform of the matrix $A^{\mathscr{D}}A (\|y\| = 1, \ y \in \text{Im}(A^{\mathscr{D}}A))$. Since the sphere is closed, and the function $\varphi(y)$ is known to be continuous on it, then the numbers $\mu = \min_{y \in K} \varphi(y)$ and $\lambda = \max_{y \in K} \varphi(y)$ are determined and, as a consequence of the function $\varphi(y)$ (on K), the inequalities $\mu > 0$ and $\lambda > 0$ are valid. But then for all $x \in \text{Im}(A^{\mathscr{D}}A)$, including the solution of the system $x' = Ax$, $x(0) = A^{\mathscr{D}}A\alpha$, the inequalities

$$\mu(x, x) \leqslant (Yx, x) \leqslant \lambda(x, x)$$

hold, from which it follows that

$$u \leqslant \lambda \|x\|^2, \qquad -\|x\|^2 \leqslant -\lambda^{-1}u, \qquad \|x\|^2 \leqslant \mu^{-1}u. \quad (2.7.13)$$

Now, taking (2.7.12) and (2.7.13) into account, it is easy to arrive at the inequalities

$$u' \leqslant -\lambda^{-1}u, \qquad u(t) \leqslant e^{-t/\lambda}u(0),$$

$$\|x(t)\|^2 \leqslant \mu^{-1}u(t) \leqslant \mu^{-1}e^{-t/\lambda}u(0) \leqslant \lambda\mu^{-1}e^{-t/\lambda}\|x(0)\|^2,$$

from which it follows that $x(t) \to 0$ when $t \to \infty$, i.e. when $t \to \infty$

$$x(t) = e^{At} A^{\mathscr{D}}A\alpha \to 0$$

for an arbitrary vector α. This means that when $t \to \infty$ the limiting relationship (2.7.9) is valid. Using this and formula (2.7.8) one can prove that all nonzero eigen-numbers of the matrix A lie in the left-hand semi-plane (we leave the proof to the reader).

A consequence of the proved lemma is

Theorem 2.7.4 For the matrix A to be stable, it is necessary and sufficient that its index does not exceed unity and equation (2.7.6) has the solution Y, which is the Hermitian matrix, positively defined on the transform of the matrix $A^{\mathscr{D}}A$.

Remark 2.7.2 Sometimes the matrix Y, which is the solution of equation (2.7.6), can be written explicitly: if, for example, the matrix A is normal, i.e. in it $A^*A = AA^*$, then in the case of stability of the matrix A the solution (2.7.10) of equation (2.7.6) has the form

$$Y = (A^{\mathscr{D}}A)^* \int_0^\infty e^{(A^*+A)t}\, dt \cdot A^{\mathscr{D}}A. \tag{2.7.14}$$

Therefore, by calculating the integral in (2.7.14), we obtain the explicit expression for the matrix Y.

In order to calculate the integral in (2.7.14), one can make use of the easily verifiable formula which is valid for any matrix M:

$$\int e^{tM}\, dt = M^{\mathscr{D}} e^{Mt} + (E - M^{\mathscr{D}}M)\sum_{i=1}^{k} \frac{1}{i!} M^{i-1} t^i + C, \tag{2.7.15}$$

where k is the index of the matrix M, and C is an arbitrary matrix. In this case it should be taken into consideration that for a normal matrix, the equalities

$$(A^{\mathscr{D}}A)^*[E - (A^* + A)^{\mathscr{D}}(A^* + A)] = 0, \tag{2.7.16}$$

$$(A^{\mathscr{D}}A)^*(A^* + A)^{\mathscr{D}} = (A^* + A)^{\mathscr{D}} \tag{2.7.17}$$

hold. This is most easily proved by transition to the diagonal form with the help of a unitary matrix.

Taking (2.7.15) and (2.7.16) into account we obtain:

$$\int_0^\infty e^{t(A^*+A)}\, dt = -(A^* + A)^{\mathscr{D}}.$$

But then, as a consequence of (2.7.14) and (2.7.17) it appears that

$$Y = -(A^* + A)^{\mathscr{D}}. \tag{2.7.18}$$

Note also that the obtained formula (2.7.18) opens the way to an approximate representation of the matrix Y because, for the matrix (2.7.18), the approximation

$$[-(A^* + A)^2 + \tau^2 E]^{-1}(A^* + A) \sim -(A^* + A)^{\mathscr{D}}$$

is valid. This idea forms the basis for the formulation of the following theorem about indices.

Theorem 2.7.5 The index of the matrix A equals k if and only if the matrix

$$(A^k + \tau^2 E)^{-1} A^k \qquad (2.7.19)$$

has a finite limit when $\tau \to 0$.

The proof can be carried out through a transition from the matrix (2.7.19) to its canonical form (the proof is left to the reader).

2.8. Application of the resolving pair of matrices

Let us consider the problem

$$A(t)x'(t) = B(t)x(t) + f(t), \quad \alpha \leqslant t \leqslant \beta, \qquad (2.8.1)$$

$$x(\alpha) = a, \qquad (2.8.2)$$

where A and B are $(m \times n)$ matrices, and let us assume that the columns of the matrix $B(t)$ are linearly independent for all $t \in [\alpha, \beta]$ (the pair of matrices (B, A) is perfect) and the problem (2.7.1), (2.7.2) has a unique solution. With such an assumption, as in [1], one can conclude that if the problem

$$B^+ A x' = x + B^+ f, \qquad (2.8.3)$$

$$x(\alpha) = a, \qquad (2.8.4)$$

also has a unique solution, then the problems (2.8.1), (2.8.2) and (2.8.3), (2.8.4) are equivalent, and the solution of the initial problem (2.8.1), (2.8.2) is reduced to solving the problem (2.8.3), (2.8.4), and on the basis of Corollary 1.2.1, equation (2.8.3) can also be written as

$$B^* A x' = B^* B x + B^* f.$$

It is obvious that if the matrix $B^+ A = (B^* B)^{-1} B^* A$ either has the property Ω, or one of the inequalities $\mathrm{ind}(B^+ A) \leqslant 1$, $\mathrm{ind}(B^+ A) \leqslant 2$ is satisfied for it, then for the solution of problem (2.8.3), (2.8.4), one can make use of the corresponding difference schemes developed in Sections 2.2–2.6. However, such an approach to solving the problem (2.8.1), (2.8.2) has a disadvantage, namely it rules out the possibility of verifying the matching condition of the initial data with the right-hand side because this condition, which emerges after equation (2.8.1) has been multiplied by the matrix $(E - BB^+)$, has the form

$$(E - BB^+) A x'|_{t=\alpha} = (E - BB^+) f|_{t=\alpha} \qquad (2.8.5)$$

not containing $x(\alpha) = a$.

Another approach to the problem (2.8.1), (2.8.2) suggests using the semi-inverse matrix Y from the resolving pair of matrices (A^B, Y). With such an approach, instead of the problem (2.8.3), (2.8.4) the problem

$$Y A x' = x + Y f, \qquad (2.8.6)$$

$$x(\alpha) = a, \qquad (2.8.7)$$

is obtained, and the matching condition (2.8.5), as will be shown in the next chapter under the same assumptions, is written explicitly and does not contain any unknown solutions.

However, to reduce the problem (2.8.1), (2.8.2) to the equivalent problem (2.8.6), (2.8.7), it is necessary to know the matrix Y, which is difficult to find not least because the algorithm to determine it, developed in Section 1.11, is complicated. In this connection, it is appropriate to separate classes of pairs of matrices, for which the above-mentioned algorithm rapidly leads to the solution. Let us consider, for example, the pair of matrices (A, B) for which with a certain choice of semi-inverse matrices, the equality $(E - BB^-)A_1 = 0$ is satisfied or, in greater detail, the equality

$$(A - BB^- A)B^- A[E - (A - BB^- A)^- (A - BB^- A)] = 0$$

holds (see the algorithm at the end of Section 1.11). For any semi-inverse matrices, this equality holds if the equality

$$(E - BB^-)A = 0 \tag{2.8.8}$$

is valid, which means that the pair of matrices (B, A) is quite perfect or if

$$A[E - (A - BB^- A)^- (A - BB^- A)] = 0, \tag{2.8.9}$$

which by virtue of Lemma 1.4.4 is equivalent to the equality $\operatorname{Im} A \cap \operatorname{Im} B = 0$ being satisfied. In the first case (see (2.8.8)), according to our algorithm, as the matrix Y one can take any semi-inverse matrix B^- to the matrix B (including the matrix B^+). In the second case, as a consequence of the fact that the equality (2.8.9) is satisfied, we have $A_1 = 0$ and $M_1 = A$, and because $V_0 = 0$ and $G_0 = E$, we obtain for V_i the value

$$V_1 = A^- AB^- A(A - BB^- A)^- (E - BB^-).$$

Finally, in view of the fact that the columns of the matrix B are linearly independent and, consequently, the third term in formula (1.11.38) is zero, we arrive at the following expression for the matrix Y:

$$Y = B^- - A^- AB^- A(A - BB^- A)^- (E - BB^-) \tag{2.8.10}$$

(the semi-inverse matrices in (2.8.10) are arbitrary ones).

It is interesting to note that the matrix YA involved in equation (2.8.6), as a consequence of the equality (2.8.9), has a simple form:

$$YA = (E - A^- A)B^- A. \tag{2.8.11}$$

Moreover, $(YA)^2 = 0$, i.e. the matrix YA is nilpotent, and its index does not exceed two. In this case it appears that, generally speaking, $YA \neq 0$. In fact, let

$$A = \begin{pmatrix} 0 & 0 \\ 0 & 0 \\ 1 & 1 \end{pmatrix}, \qquad B = \begin{pmatrix} 1 & 0 \\ 0 & 1 \\ 0 & 0 \end{pmatrix},$$

then, by taking as semi-inverse matrices in (2.8.11) the matrices

$$A^- = \begin{pmatrix} 0 & 0 & 1 \\ 0 & 0 & 0 \end{pmatrix}, \quad B^- = \begin{pmatrix} 1 & 0 & 0 \\ 0 & 1 & 1 \end{pmatrix},$$

we obtain:

$$YA = \begin{pmatrix} -1 & -1 \\ 1 & 1 \end{pmatrix}.$$

If, however,

$$B^- = \begin{pmatrix} 1 & 0 & 0 \\ 0 & 1 & 0 \end{pmatrix},$$

then $B^- A = 0$, and therefore $YA = 0$. In any event, there emerges the problem of solving the equation of the form

$$Ax' = x + f, \tag{2.8.12}$$

in which $A^2 = 0$.

To solve this problem, we multiply equation (2.8.12) by the matrix A. As a result, we obtain:

$$Ax = -Af. \tag{2.8.13}$$

At the same time we write equation (2.8.12) in the equivalent form,

$$(Ax)' = (E + A')x + f, \tag{2.8.14}$$

and substitute the left-hand side of equation (2.8.14) into the right-hand side of (2.8.13). This will lead to the equality

$$(E + A')x = -f - (Af)' = -(E + A')f - Af'.$$

Thus, any solution of equation (2.8.12) satisfies the algebraic system

$$Ax = -Af, \tag{2.8.15}$$

$$(E + A')x = -(E + A')f - Af'. \tag{2.8.16}$$

The opposite is also true: any solution of the algebraic system (2.8.15), (2.8.16) is the solution of equation (2.8.12) (we leave the proof to the reader).

A general solution of the system (2.8.15), (2.8.16) and its existence conditions can be obtained, for example, with the help of semi-inverse matrices, which are dealt with in Lemma 1.3.6. In this chapter, however, we are interested in the case when the system (2.8.15), (2.8.16) (or, equivalently, equation (2.8.12)) has only one solution; therefore it should be assumed that the matrix columns of the system (2.8.15), (2.8.16) are linearly independent. With such an assumption the semi-inverse matrix $M^- = (M^*M)^- M^*$ transforms into the pseudo-inverse matrix $M^+ = (M*M)^{-1}M*$ (see Corollary 1.2.1), and one can then formulate the following theorem.

Theorem 2.8.1 In equation (2.8.12) let the matrix A satisfy the condition $A^2 = 0$. Then, for the solution of equation (2.8.12) to exist and to be unique, it is necessary and sufficient that the matrix

$$C = [A*A + (E + A')*(E + A')]^{-1} \qquad (2.8.17)$$

exists and the equalities

$$AC(E + A')*Af' = 0, \qquad (2.8.18)$$

$$(E + A')C(E + A')*Af' = Af'. \qquad (2.8.19)$$

are satisfied. In this case, if the solution of equation (2.8.12) exists, then it is expressed by the formula

$$x = -f - C(E + A')*Af'. \qquad (2.8.20)$$

Proof We represent the solution of the system (2.8.15), (2.8.16) as

$$x = -f - y. \qquad (2.8.21)$$

Then for y we obtain the system

$$Ay = 0 \qquad (E + A')y = Af', \qquad (2.8.22)$$

which is known to have a unique solution if and only if the matrix columns of the system are linearly independent, which is equivalent to the existence of the matrix (2.8.17). If it is now taken into consideration that, on the basis of Corollary 1.2.1, the matrix

$$C \cdot (A*(E + A')*)$$

is pseudo-inverse to the matrix of the system (2.8.22), then with the help of Corollary 1.3.1 it is easy to obtain the existence conditions (2.8.18) and (2.8.19) and the unique solution of the system (2.8.22) which has the form

$$y = C(E + A')*Af'.$$

Substituting this into (2.8.21) leads to formula (2.8.20), which completes the proof of the theorem.

If Theorem 2.8.1 is applied to equation (2.8.6), taking into account formulae (2.8.10) and (2.8.11), we obtain the solution of the problem (2.8.6), (2.8.7) (and, consequently, by virtue of equivalency the solution of the initial problem (2.8.1), (2.8.2)), and the matching conditions of the initial data (2.8.2) with the right-hand side f become obvious. If, moreover, the semi-inverse matrices in (2.8.10) and (2.8.11) are replaced with pseudo-inverse ones and, after that, Theorems 1.2.2 and 1.2.3 are used, then we arrive at the formula for an approximate solution of the problem (2.8.1), (2.8.2) not containing semi-inverse matrices.

To conclude, we give one interesting example.
In the system (2.8.12) let

$$A(t) = t^3 \begin{pmatrix} \frac{1}{2}\sin\frac{2}{t} & -\sin^2\frac{1}{t} \\ \cos^2\frac{1}{t} & -\frac{1}{2}\sin\frac{2}{t} \end{pmatrix}, \quad 0 \leqslant t < 1. \qquad (2.8.23)$$

It is easy to verify that $A^2(t) \equiv 0$ and the matrix $E + A'(t)$ for all $t \in [0,1)$ is nonsingular. But then the matrix (2.8.17) also is obviously nonsingular, and to solve the system (2.8.12) with the matrix (2.8.23), one can apply Theorem 2.8.1. In this case, however, one can proceed in a much simpler way because, under the assumption of nonsingularity of the matrix $E + A'$, the following theorem is valid.

Theorem 2.8.2 In the system (2.8.12) let the matrix A satisfy the conditions $A^2 = 0$ and let the matrix $E + A'$ be nonsingular. Then the solution of equation (2.8.12) exists, is unique, and is expressed by the formula

$$x = -f - (E + A')^{-1}Af'. \qquad (2.8.24)$$

Proof The fact that the solution of the system (2.8.12) (if it exists) is unique and is expressed by formula (2.8.24), immediately follows from the nonsingularity of the matrix $E + A'$ because in such a case the second equation of (2.8.22) is uniquely solvable and substitution of the solution of this equation into the equality (2.8.21) leads to formula (2.8.24).

In order to make sure that the vector (2.8.24) is indeed the solution of the system (2.8.12) for any vector f, it is sufficient to show that the second summand of (2.8.24) satisfies the first equation of (2.8.22). But this easily follows from the equalities $A^2 = 0$ and $AA' = -A'A$, the second of which is the consequence of the first.

In fact, since the matrix $(E + A')^{-1}$ is a polynomial of the matrix A', then as a consequence of the equality $AA' = -A'A$, there exists a matrix C such that $A(E + A')^{-1} = CA$. But then $A(E + A')^{-1}A = CA^2 = 0$ and the first equality of (2.8.22) is satisfied identically with respect to f. The theorem is thereby proved.

By applying the proved theorem to solve the system (2.8.12) with the matrix (2.8.23), we obtain:

$$x(t) = -f(t) - \frac{1}{1+t}A(t)f'(t). \qquad (2.8.25)$$

It will scarcely be possible to obtain a continuous solution of (2.8.25) with the help of long-standing methods associated with elementary transformations, including that of reducing the matrices to a canonical form.

2.9. The boundary layer of errors. Stability of linear combinations of the components of a difference solution

A direct extension of the results of the theory of difference schemes to the case of singular systems is, generally speaking, impossible. For the implicit Euler scheme

$$(A_i - \tau B_i)\frac{x_i - x_{i-1}}{\tau} = B_i x_{i-1} + f_i, \quad i = 1, \ldots, N; \qquad (2.9.1)$$

$$N\tau = \beta - \alpha, \qquad x_0 = a, \qquad (2.9.2)$$

which approximates the problem

$$A(t)x'(t) = B(t)x(t) + f(t), \quad \alpha \leqslant t \leqslant \beta, \qquad (2.9.3)$$

$$x(\alpha) = a, \qquad (2.9.4)$$

it is already impossible, without additional assumptions about the matrices (2.9.1) and (2.9.3), to write the estimate

$$\|x_i\|_1 \leqslant M\|x_0\|_2 + P\|f\|_3$$

that determines the stability and convergence of the scheme (2.9.1), (2.9.2).

Ultimately, this is associated with that property of the scheme, the presence of which gives rise to the so-called 'boundary layer of errors', which sometimes extends even to the entire portion of integration.

In order to understand the reasons for such a phenomenon, we consider a number of examples.

Example 2.9.1 We apply the implicit Euler Scheme (2.9.1), (2.9.2) to solve the scalar problem

$$0 \cdot x' = x - \varphi, \qquad x(0) = a. \qquad (2.9.5)$$

It is obvious that its accurate solution is $x = \varphi$; therefore the matching condition $\varphi(0) = a$ must be satisfied. The difference scheme (2.9.1), (2.9.2) in this case has the form

$$-\tau \frac{x_i - x_{i-1}}{\tau} = x_{i-1} - \varphi_i, \qquad x_0 = a, \quad i = 1, \ldots, N,$$

and, as can easily be understood, when $i = 1, \ldots, N$ it gives an accurate solution $x_i = \varphi_i$, such that an error can arise only at the zero point as a result of incorrectly (not matched with the right-hand side φ) taking the initial data $x_0 \neq \varphi(0)$. 'The boundary layer of errors' degenerates in this case into one point.

Example 2.9.2 In (2.9.3) and (2.9.4) let

$$A = \begin{Bmatrix} 0 & 1 \\ 0 & 0 \end{Bmatrix}, \qquad B = E, \qquad f = \begin{Bmatrix} 0 \\ -\varphi \end{Bmatrix}.$$

The accurate solution of equation (2.9.3) in this case is also unique and is expressed by the formulae

$$x(t) = \begin{pmatrix} x_1(t) \\ x_2(t) \end{pmatrix}, \qquad x_1 = \varphi', \qquad x_2 = \varphi, \qquad (2.9.6)$$

from which follows the matching condition of initial data with the right-hand side:

$$x_1(0) = \varphi'(0), \qquad x_2(0) = \varphi(0). \qquad (2.9.7)$$

As far as the difference condition is concerned, it is calculated from the formula

$$x_i = Cx_{i-1} + \tau D f_i, \quad i = 1, \dots, N, \qquad (2.9.8)$$

where

$$C = (A - \tau E)^{-1} A, \qquad D = (A - \tau E)^{-1}.$$

In this case, as a consequence of the nilpotence of the matrix C, i.e. as a consequence of the fact that $C^2 = 0$, formula (2.9.8) can also be written as

$$x_1 = Cx_0 + \tau D f_1, \qquad (2.9.9)$$

$$x_i = \tau C D f_{i-1} + \tau D f_i, \quad i = 2, \dots, N. \qquad (2.9.10)$$

Formula (2.9.9) for the vector x_1 yields

$$x_1 = \begin{pmatrix} \dfrac{\varphi_1 - x_{2,0}}{\tau} \\ \varphi_1 \end{pmatrix}, \qquad (2.9.11)$$

where $\varphi_1 = \varphi(\tau)$ and $x_{2,0}$ is the second component of the initial vector $x_0 = a$.

A comparison of (2.9.11) with (2.9.6) shows that, if $x_{2,0} \neq \varphi(0)$, i.e. $x_{2,0}$ is not matched with the right-hand side, then at the first point of the difference grid $(i = 1)$ in the first component of the vector x_1, an error of order $O(\tau^{-1})$ is possible, and, at the same time, the second component of the vector x_1 is accurately calculated.

At the other points of the grid (according to formula (2.9.10)), we have

$$x_i = \begin{pmatrix} \dfrac{\varphi(i\tau) - \varphi((i-1)\tau)}{\tau} \\ \varphi(i\tau) \end{pmatrix}$$

and, consequently, at the grid points $i = 2, \dots, N$ the difference scheme (2.9.1), (2.9.2) gives an approximate value of the accurate solution (2.9.6) (with an error of order $O(\tau)$ in the first component, but the second component is accurately calculated). Thus, the 'boundary layer of errors' in this case covers two points $(i = 0, 1)$.

Example 2.9.3 In (2.9.3) and (2.9.4) let

$$A = \begin{pmatrix} 0 & 1 & 0 \\ 0 & 0 & 1 \\ 0 & 0 & 0 \end{pmatrix}, \qquad B = E, \qquad f = \begin{pmatrix} 0 \\ 0 \\ -\varphi \end{pmatrix}.$$

An accurate solution of the problem (2.9.3), (2.9.4) and the matching conditions in this case have the form

$$x = \begin{pmatrix} \varphi'' \\ \varphi' \\ \varphi \end{pmatrix},$$

$$x_1(0) = \varphi''(0), \qquad x_2(0) = \varphi'(0), \qquad x_3 = \varphi(0). \tag{2.9.12}$$

In view of the fact that $C^3 = 0$, formula (2.9.8) is calculated in the following way:

$$x_1 = Cx_0 + \tau D f_1, \tag{2.9.13}$$

$$x_2 = C^2 x_0 + \tau CD f_1 + \tau D f_2, \tag{2.9.14}$$

$$x_i = \tau C^2 D f_{i-2} + \tau CD f_{i-1} + \tau D f_i, \quad i = 3, \ldots, N. \tag{2.9.15}$$

Through a direct calculation using formulae (2.9.13)–(2.9.15) we make sure that

$$x_1 = \begin{pmatrix} \dfrac{\varphi(\tau) - x_{3,0}}{\tau^2} - \dfrac{x_{2,0}}{\tau} \\ \dfrac{\varphi(\tau) - x_{3,0}}{\tau} \\ \varphi(\tau) \end{pmatrix}, \qquad x_2 = \begin{pmatrix} \dfrac{\varphi(2\tau) - 2\varphi(\tau) + x_{3,0}}{\tau^2} \\ \dfrac{\varphi(2\tau) - \varphi(\tau)}{\tau} \\ \varphi(2\tau) \end{pmatrix}, \tag{2.9.16}$$

where $x_{2,0}$ and $x_{3,0}$ are the second and third components of the initial vector $x_0 = a$, and

$$x_i = \begin{pmatrix} \dfrac{\varphi(i\tau) - 2\varphi((i-1)\tau) + \varphi((i-2)\tau)}{\tau^2} \\ \dfrac{\varphi(i\tau) - \varphi((i-1)\tau)}{\tau} \\ \varphi(i\tau) \end{pmatrix}, \qquad i = 3, \ldots, N. \tag{2.9.17}$$

A comparison of (2.9.16) with (2.9.12) shows that if $x_{2,0} \neq \varphi'(0)$ and $x_{3,0} \neq \varphi(0)$, then when $i = 1, 2$, an error of order $O(\tau^{-2})$ is possible. But even if $x_{2,0} = \varphi'(0)$ and $x_{3,0} = \varphi(0)$, i.e. the initial data are matched with the right-hand side, the error in the first component of the vector x_1 is not excluded: it can have the order $O(1)$.

In the other grid points (when $i = 3, \ldots, N$), however, as shown by a comparison of (2.9.17) with (2.9.12), the difference scheme gives an approximate solution of the problem (2.9.3), (2.9.4). As far as the third component of the solution is concerned, at all points of the grid $(i = 1, \ldots, N)$ it is calculated accurately. The 'boundary layer of errors' here covers three points: $i = 0, 1, 2$.

Example 2.9.4   In (2.9.3) and (2.9.4) let

$$
A = \begin{pmatrix} 0 & 2-e^t & 1-e^t & e^t-1 \\ 0 & -2 & -1 & 2 \\ 0 & 1 & 1 & -1 \\ 0 & -2 & -1 & 2 \end{pmatrix}, \quad f = \begin{pmatrix} e^t - e^{2t} \\ -2e^t \\ 0 \\ -2e^t \end{pmatrix}, \quad (2.9.18)
$$

$$
x(0) = (1\ 1\ 1\ 1)^T.
$$

Then, the accurate solution of the problem (2.9.3), (2.9.4) will be the vector $x(t) = (e^t e^t e^t e^t)^T$.

We suggest that the reader will by himself make sure that the first component of the difference solution obtained according to the scheme (2.9.1), (2.9.2), with the initial data (2.9.18), 'goes wrong', and the other components are calculated with an error of order $O(\tau)$. The 'boundary layer of errors' in this example is 'smeared out' for the entire portion of the integration.

The examples considered above suggest that the implicit Euler scheme, even in the case when it is accompanied by the 'boundary layer of errors', is not useless: some components of the solution, or, more generally, some linear combinations of its components, can be quite well calculated. In this connection the problem arises of searching for such combinations and establishing signatures indicating that the combinations are such.

Results which open up some possibilities here, can be obtained using the method reported by Samarsky, who obtained, based on this method, simple signatures of stability of difference schemes of mathematical physics [21, 22].

So, let us examine the equation

$$
A_i \frac{x_i - x_{i-1}}{\tau} = B_i x_{i-1}, \quad i = 1, 2, \ldots, \tau > 0. \quad (2.9.19)
$$

For compactness, we write it as

$$
A_i y_i = B_i x_{i-1}, \quad i = 1, 2, \ldots, \quad (2.9.20)
$$

where

$$
y_i = \frac{x_i - x_{i-1}}{\tau}. \quad (2.9.21)
$$

We multiply equation (2.9.20) by some matrix M_i, and then multiply scalarly by the vector $2\tau y_i$. As a result, we arrive at the equality

$$
2\tau(M_i A_i y_i, y_i) = 2\tau(M_i B_i x_{i-1}, y_i),
$$

from which, obviously, it follows that

$$2\tau \operatorname{Re}(M_iA_iy_i, y_i) = 2\tau \operatorname{Re}(M_iB_ix_{i-1}, y_i). \tag{2.9.22}$$

Suppose now that $(M_iB_i)^* = M_iB_i$. Then, in view of (2.9.21), we obtain:

$$
\begin{aligned}
2\tau \operatorname{Re}(M_iB_ix_{i-1}, y_i) &= \tau[(M_iB_ix_{i-1}, y_i) + (y_i, M_iB_ix_{i-1})] \\
&= \tau[(M_iB_ix_{i-1}, y_i) + (M_iB_iy_i, x_{i-1})] \\
&= (M_iB_ix_i, x_i) - (M_iB_ix_{i-1}, x_{i-1}) - \tau^2(M_iB_iy_i, y_i).
\end{aligned}
$$

Moreover,

$$
\begin{aligned}
2\tau \operatorname{Re}(M_iA_iy_i, y_i) &= \tau[(M_iA_iy_i, y_i) + (y_i, M_iA_iy_i)] \\
&= \tau[(M_iA_iy_i, y_i) + (A_i^*M_i^*y_i, y_i)] \\
&= \tau((A_i^*M_i^* + M_iA_i)y_i, y_i).
\end{aligned}
$$

But then the equality (2.9.22) takes on the form

$$- \tau((A_i^*M_i^* + M_iA_i + \tau M_iB_i)y_i, y_i) + (M_iB_ix_i, x_i) = (M_iB_ix_{i-1}, x_{i-1}). \tag{2.9.23}$$

We now note that, if the first Hermitian in the left-hand side of the equality (2.9.23) is negatively semi-determined, then

$$(M_iB_ix_i, x_i) \leqslant (M_iB_ix_{i-1}, x_{i-1}).$$

If, moreover, the matrix MB is constant, then for all $i = 1, 2, \ldots,$ the inequality

$$(MBx_i, x_i) \leqslant (MBx_{i-1}, x_{i-1})$$

holds and, consequently, for any $k \leqslant s$,

$$(MBx_k, x_k) \leqslant (MBx_s, x_s). \tag{2.9.24}$$

Hence the following lemma is proved.

Lemma 2.9.1 If the matrices $A_i, B_i,$ and M_i $(i = 1, 2, \ldots)$ are such that

$$A_i^*M_i^* + M_iA_i + \tau M_iB_i \leqslant 0, \quad (M_iB_i)^* = M_iB_i,$$

and the matrix M_iB_i is constant (does not depend on i), then when $k \geqslant s$, for all solutions of equation (2.9.19), the estimate (2.9.24) is valid.

Theorem 2.9.1 If the matrices $A_i, B_i,$ and M_i $(i = 1, 2, \ldots)$ and the real number $\rho \neq 0$ are such that

$$A_i^*M_i^* + \frac{1}{\rho}M_iA_i + \frac{\tau}{\rho}M_iB_i \leqslant 0,$$

and the matrix

$$D = M_i\left(\frac{1}{\rho}B_i + \frac{1-\rho}{\tau\rho}A_i\right)$$

is self-conjugate $(D^* = D)$ and constant, then when $k \geqslant s$, for all solutions of equation (2.9.19), the estimate

$$(Dx_k, x_k) \leqslant \rho^{2(k-s)}(Dx_s, x_s)$$

holds.

Proof Let us make the substitution $x_i = \rho^i y_i$ in equation (2.9.19). Then, for the vector y, we obtain the equation

$$A_i \frac{y_i - y_{i-1}}{\tau} = \left(\frac{1}{\rho} B_i + \frac{1-\rho}{\tau\rho} A_i \right) y_{i-1}.$$

By applying to it Lemma 2.9.1 and then making the inverse substitution $y_i = \rho^{-i} x_i$, we arrive at the result of Theorem 2.9.1.

We now limit ourselves to considering difference equations of the form

$$A_i \frac{x_i - x_{i-1}}{\tau} = x_{i-1}, \quad i = 1, 2, \dots. \tag{2.9.25}$$

However, a word of caution is in order here: one must not think that the matrix A_i in (2.9.25) is necessarily the approximation of the matrix $A(t)$ from the equation

$$A(t)x'(t) = x(t).$$

The matrix A_i can, indeed, be some approximation of the matrix $A(t)$ at the point $t = \tau i$ (see, for example, (2.9.1) for $B = E$), but it can also be a matrix obtainable when the difference equations examined in Sections 2.4–2.6 are brought to the form (2.9.25).

Let us prove the following lemma.

Lemma 2.9.2 Let the matrices A_i and C_i $(i = 1, 2, \dots)$ be such that the matrices A_i and $A_i + \tau E$ are nonsingular for all i, and the matrix $C_i^* C_i$ is constant (does not depend on i). Then, for all the solutions of equation (2.9.25), when $k \geqslant s$, to satisfy the estimate

$$\| C_k x_k \| \leqslant \| C_s x_s \|, \tag{2.9.26}$$

it is necessary and sufficient that for all i the inequality

$$A_i^* C_i^* C_i + C_i^* C_i A_i + \tau C_i^* C_i \leqslant 0 \tag{2.9.27}$$

be satisfied (for sufficiency of the condition (2.9.27), nonsingularity of the matrices A_i and $A_i + \tau E$ is not required).

Necessity It is easy to see that, to solve equation (2.9.25), the relationship $x_i = S_i x_{i-1}$ is valid, where

$$S_i = A_i^{-1}(A_i + \tau E),$$

and, by virtue of the assumptions made, the matrix S_i is nonsingular and, consequently, one can write

$$x_0 = (S_{i-1} \cdots S_1)^{-1} x_{i-1},$$

from which it is clear that if x_0 is arbitrary, then x_{i-1} also is arbitrary. In view of this remark, as well as of the inequality (2.9.26), we find that at an arbitrary x_0 and, hence, also at x_{i-1}, the inequality

$$(C_i^* C_i S_i x_{i-1}, S x_{i-1}) \leqslant (C_i^* C_i x_{i-1}, x_{i-1})$$

holds, which, by introducing the vector $y_i = A_i^{-1} x_{i-1}$, can be rewritten also as

$$(C_i^* C_i (A_i + \tau E) y_i, (A_i + \tau E) y_i) \leqslant (C_i^* C_i A_i y_i, A_i y_i),$$

and the vector y_i also is, obviously, arbitrary.

From the inequality obtained it follows that

$$((A_i^* C_i^* C_i + C_i^* C_i A_i + \tau C_i^* C_i) y_i, y_i) \leqslant 0,$$

and since the vector y_i is arbitrary, the inequality (2.9.27) is proved.

The sufficiency follows immediately from Lemma 2.9.1 (one must put $M = C^* C$ and $B = E$).

Theorem 2.9.2 Let the matrices A_i and C_i $(i = 1, 2, \ldots)$ and the real number $\rho \neq 0$ be such that the matrices $A_i, E + [(1 - \rho)/\tau] A_i$, and $A_i + \tau E$ are nonsingular, and the matrix C_i^* is constant. Then, for all the solutions of equation (2.9.25) when $k \geqslant s$ to satisfy the estimate

$$\| C_k x_k \| \leqslant |\rho|^{k-s} \| C_s x_s \|, \tag{2.9.28}$$

it is necessary and sufficient that for all i the inequality

$$A_i^* C_i^* C_i + C_i^* C_i A_i + \tau C_i^* C_i + \frac{1 - \rho^2}{\tau} A_i^* C_i^* C_i A_i \leqslant 0 \tag{2.9.29}$$

be satisfied (for sufficiency of the condition (2.9.29), nonsingularity of the matrices A_i and $A_i + \tau E$ is not needed).

Proof By making the substitution $x_i = \rho^i y_i$ in equation (2.9.25), we obtain:

$$A_i \frac{y_i - y_{i-1}}{\tau} = \left\{ \frac{1}{\rho} E + \frac{1 - \rho}{\tau \rho} A_i \right\} y_{i-1}. \tag{2.9.30}$$

Furthermore, it appears that under the assumptions of the theorem, equation (2.9.30) can be brought to the form (2.9.25) and Lemma 2.9.2 can be applied, after which the inverse replacement $y_i = \rho^{-i} x_i$ leads to the result of Theorem 2.9.2.

It is obvious that the estimates (2.9.26) and (2.9.28) indicate the degree of stability of linear combinations $y_i = C_i x_i$ of the components of the difference solution. Later, in accordance with terminology used by A. A. Samarsky, the linear combinations $y_i = C_i x_i$ will be called stable if the estimate (2.9.26) is satisfied for them, and ρ-stable if the estimate (2.9.28) is satisfied (it is clear that when $\rho = 1$, the ρ-stability coincides with stability).

Let us now consider the inhomogeneous difference problem

$$A_i \frac{x_i - x_{i-1}}{\tau} = x_{i-1} + f_i, \quad i = 1, \ldots, N; \qquad N\tau = \beta - \alpha, \qquad (2.9.31)$$

$$x_0 = a.$$

The solution of this problem is represented as

$$x_i = S_{i,1} x_0 + \tau \sum_{k=1}^{i} S_{i,k+1} A_k^{-1} f_k, \qquad (2.9.32)$$

where

$$S_{i,k} = T_i T_{i-1} \cdots T_k, \quad i \geqslant k, \qquad S_{i,i+1} = E,$$

$$T_j = A_j^{-1}(A_j + \tau E), \quad j = 1, 2, \ldots,$$

i.e. $S_{i,k}$ is the transition operator of the difference solution from the point $k - 1$ to the point i.

Suppose that the linear combinations $y_i = C_i x_i$ of the components of the solution of problem (2.9.31) are ρ-stable. Then by multiplying the equality (2.9.32) by the matrix C_i, according to the estimate (2.9.28), we obtain:

$$\| C_i x_i \| \leqslant |\rho|^i \| C_0 x_0 \| + \tau \sum_{k=1}^{i} |\rho|^{i-k} \| C_k A_k^{-1} f_k \|. \qquad (2.9.33)$$

Thus, the following theorem is proved.

Theorem 2.9.3 Let the matrices A_i and C_i $(i = 1, 2, \ldots)$ and the real number $\rho \neq 0$ be such that the matrices $\tau E + (1 - \rho) A_i$ and A_i are nonsingular, the matrix $C_i^* C_i$ is constant (does not depend on i), and the inequality (2.9.29) is satisfied. Then, for the solution of the problem (2.9.31), the estimate (2.9.33) is valid.

The estimate (2.9.33) makes it possible to obtain results for the convergence of the difference linear combinations to accurate ones. Let us consider, for example, the implicit Euler scheme

$$(A_i - \tau E) \frac{x_i - x_{i-1}}{\tau} = x_{i-1} + f_i, \quad i = 1, \ldots, N; \qquad N\tau = \beta - \alpha, \quad (2.9.34)$$

$$x_0 = a, \qquad (2.9.35)$$

which approximates to the problem

$$A(t)x'(t) = x(t) + f(t), \quad \alpha \leqslant t \leqslant \beta, \qquad (2.9.36)$$

$$x(\alpha) = a. \qquad (2.9.37)$$

The role of the matrix A_k in (2.9.33) is played by the matrix $A_k - \tau E$. In this case the inequality (2.9.29) that defines the ρ-stability assumes the form

$$A_i^* C_i^* C_i + C_i^* C_i A_i - \tau C_i^* C_i + \frac{1 - \rho^2}{\tau \rho^2} A_i^* C_i^* C_i A_i \leqslant 0. \qquad (2.9.38)$$

Thus, with regard to the problem (2.9.34), (2.9.35), it is possible to formulate the corollary that follows from Theorem 2.9.3.

Corollary 2.9.1 Let the matrices A_i and C_i, and the real number $\rho \neq 0$, be such that the matrices $A_i - \tau E$ and $\rho \tau E + (1 - \rho) A_i$, $i = 1, 2$, are nonsingular, the matrix C^*C is constant, and the inequality (2.9.38) is satisfied. Then, for the solution of the problem (2.9.34), (2.9.35), the estimate

$$\| C_i x_i \| \leqslant |\rho|^i \| C_0 x_0 \| + \tau \sum_{k=1}^{i} |\rho|^{i-k} \| C_k (A_k - \tau E)^{-1} f_k \| \qquad (2.9.39)$$

is valid.

In order to prove the convergence of $C_i x_i$ to $C(i\tau) x(i\tau)$, we introduce into our consideration the deviation $v_i = x_i - x(i\tau)$ which, as a consequence of the validity of the equalities (2.9.34)–(2.9.37) and the formula

$$x((i-1)\tau) = x(i\tau) - \tau x'(i\tau) + \frac{\tau^2}{2} x''((i-1)\theta\tau)$$

is the solution of the problem

$$(A_i - \tau E) \frac{v_i - v_{i-1}}{\tau} = v_{i-1} + \frac{\tau}{2} A_i x''((i-1)\theta\tau),$$

$$v_0 = 0, \quad i = 1, \dots, N; \qquad N\tau = \beta - \alpha,$$

and, therefore, in the case of ρ-stability, the estimate

$$\| C_i v_i \| \leqslant \frac{\tau^2}{2} \sum_{k=1}^{i} |\rho|^{i-k} \| C_k (A_k - \tau E)^{-1} A_k x''((k-1)\theta\tau) \| \qquad (2.9.40)$$

is valid (see (2.9.39)), from which in favourable cases, it follows that $\| C_i v_i \| \to 0$ when $\tau \to 0$. This can be proved by considering, for example, the problem (2.9.34), (2.9.35) with the input data of (2.9.18). With regard to this problem, it is natural to pose the question: Are the linear combinations $y = Cx$, with the matrix C of the form

$$C = \begin{pmatrix} 0 & 1 & 0 & 0 \\ 0 & 0 & 1 & 0 \\ 0 & 0 & 0 & 1 \end{pmatrix},$$

converging in it (see, for example, [4])? The answer turns out to be in the affirmative. In fact, it is easy to show that the norms in the right-hand side of the inequality (2.9.40) in this case are uniformly limited by some constant $K > 0$ and, moreover, there exists a constant $r > 0$ at which $\rho = 1 + r\tau$ satisfies the inequality (2.9.38), and the matrix $(1 + r\tau)E - rA_i$, for the formulation of Corollary 2.9.1, is nonsingular. But in such a case the inequality (2.9.40) is valid, and one can write the estimate

$$\| C_i v_i \| \leqslant \frac{\tau^2}{2} K \sum_{k=1}^{i} (1 + r\tau)^{i-k} \leqslant \frac{\tau^2}{2} K i \, e^{r(\beta - \alpha)} \leqslant \frac{\tau}{2} K \, e^{r(\beta - \alpha)},$$

i.e. $\| C_i v_i \| \to 0$ when $\tau \to 0$.

When $C = E$, Corollary 2.9.1 has a special value; it allows the investigation of the convergence of all components of the difference solution. The inequality (2.9.38) and the estimate (2.9.39) assume in this case the form

$$A_i^* + A_i - \tau E + \frac{1 - \rho^2}{\tau \rho^2} A_i^* A_i \leqslant 0, \tag{2.9.41}$$

$$\| x_i \| \leqslant |\rho|^i \| x_0 \| + \tau \sum_{k=1}^{i} |\rho|^{i-k} \| (A_k - \tau E)^{-1} f_k \|. \tag{2.9.42}$$

In particular, if $A_i^* + A_i \leqslant 0$, then when $\rho = 1$ all the assumptions of Corollary 2.9.1 are obviously satisfied, and the estimate

$$\| x_i \| \leqslant \| x_0 \| + \tau \sum_{k=1}^{i} \| (A_k - \tau E)^{-1} f_k \| \tag{2.9.43}$$

holds.

With the help of the estimates (2.9.42) and (2.9.43) one can obtain estimates for the deviations of the difference solution from an accurate one. Owing to the importance of these estimates, we shall write them as

$$\| v_i \| \leqslant \frac{\tau^2}{2} \sum_{k=1}^{i} |\rho|^{i-1} \| (A_k - \tau E)^{-1} A_k x''((k-1)\theta\tau) \|, \tag{2.9.44}$$

$$\| v_i \| \leqslant \frac{\tau^2}{2} \sum_{k=1}^{i} \| (A_k - \tau E)^{-1} A_k x''((k-1)\theta\tau) \|. \tag{2.9.45}$$

With regard to the estimate (2.9.45) we note that, since it has been derived from the assumption $A_i^* + A_i \leqslant 0$, the convergence of v_i to zero follows from it. Indeed, with such an assumption, the index of the matrix A_i does not exceed unity, and with the help of Theorem 2.7.5, one can choose a constant $L > 0$ which is independent of k and τ, for which

$$\| (A_k - \tau E)^{-1} A_k x''((k-1)\theta\tau) \| \leqslant L.$$

But then, from the estimate (2.9.45) follows the inequality

$$\| v_i \| \leqslant (\tau/2) L,$$

from which it follows that $\| v_i \| \to 0$ when $\tau \to 0$.

As far as the estimate (2.9.44) is concerned, it sometimes helps to resolve very complicated cases. Let us examine, for example, the problem

$$A(t)x'(t) = x(t) + f(t), \quad \alpha \leqslant t \leqslant \beta,$$

$$x(\alpha) = \left\{ \begin{matrix} e^{-\alpha} \\ 0 \end{matrix} \right\},$$

where

$$A(t) = \begin{pmatrix} 1 & t \\ -t^{-1} & -1 \end{pmatrix}, \qquad f(t) = \begin{pmatrix} -2e^{-t} \\ t^{-1}e^{-t} \end{pmatrix}.$$

The unique solution is the vector

$$x(t) = \begin{pmatrix} e^{-1} \\ 0 \end{pmatrix}.$$

The implicit Euler difference scheme (2.9.34), (2.9.35), as a consequence of nilpotency of the matrix A ($A^2 = 0$), is easily written in explicit form as

$$x_i = -(1/\tau)A_i x_{i-1} - (E + (1/\tau)A_i)f_i.$$

Using this equality, at the first grid point when $\alpha = 0$ we obtain:

$$x_{1,1} = e^{-\tau} + O(\tau), \qquad x_{2,1} = O(\tau^{-1}),$$

from which it is evident that when $\tau \to 0$, the error in the second component of the difference solution tends to infinity.

We leave to the reader the possibility of making sure by himself that with increasing t, the error in both components initially increases, then starts decreasing, and finally turns out in reasonable limits. An explanation for such a phenomenon can be given by analysing the inequality (2.9.41) and the estimate (2.9.44). The point here is that in the case of the problem (2.9.46), at a sufficiently small $\tau > 0$ and a sufficiently large $t > 0$, there exists a constant $K > 0$ at which the number $\rho = 1 + K\tau$ satisfies the inequality (2.9.41) and, consequently, the estimate (2.9.44) is valid. In this case it appears that the right-hand side of the estimate (2.9.44) is a value of order $O(e^{-t}/t)$. We leave the details for the reader.

We give yet another application of the inequality (2.9.41) and of the estimate (2.9.44).

Let us consider the problem

$$A(t)x'(t) = B(t)x(t) + f(t), \quad \alpha \leqslant t \leqslant \beta, \tag{2.9.46}$$

$$x(\alpha) = a,$$

in which the columns of the $(m \times n)$ matrix $B(t)$ are linearly independent, and let us assume that it has a unique solution.

Let the problem

$$B^*(t)A(t)x'(t) = B^*(t)B(t)x(t) + B^*(t)f(t), \tag{2.9.47}$$

$$x(\alpha) = a, \quad \alpha \leqslant t \leqslant \beta,$$

also have a unique solution. Then, as has already been pointed out at the beginning of the preceding section, the problems (2.9.46) and (2.9.47) are equivalent. However, the formulation of the problem (2.9.47) is, obviously, more preferable because the matrices involved in it are quadratic and, therefore, for the solution, one can apply the difference scheme of the form

$$(B_i^* A_i - \tau B_i^* B_i)\frac{x_i - x_{i-1}}{\tau} = B_i^* B_i x_{i-1} + B_i^* f_i, \qquad (2.9.48)$$

$$x_0 = a.$$

It is clear that the convergence properties of the scheme (2.9.48) will not change if they are multiplied by the inverse to the nonsingular matrix $B_i^* B_i$; therefore, instead of the scheme (2.9.48), one can study the scheme

$$(B_i^+ A_i - \tau E)\frac{x_i - x_{i-1}}{\tau} = x_{i-1} + B_i^+ f_i, \qquad (2.9.49)$$

$$x_0 = a,$$

having the form (2.9.34), (2.9.35). The inequality (2.9.41) and the estimate (2.9.44) are obtained in this case such that

$$A_i^*(B_i^*)^+ + B_i^+ A_i - \tau E + \frac{1-\rho^2}{\tau \rho^2} A_i^*(B_i^+)^* B_i^+ A_i \leqslant 0, \qquad (2.9.50)$$

$$\|v_i\| \leqslant \frac{\tau^2}{2} \sum_{k=1}^{i} |\rho|^{i-k} \|(B_k^+ A_k - \tau E)^{-1} B_k^+ A_k x''((k-1)\theta\tau)\|. \qquad (2.9.51)$$

In particular, from (2.9.50) and (2.9.51), as above, it follows that if $(B_i^+ A_i)^* + B_i^+ A_i \leqslant 0$, then the scheme (2.9.48) is converging.

To conclude, we want to discuss the possibilities of applying the matrix Y from the formulation of Lemma 2.7.1, and we notice immediately that for our purposes these possibilities are limited, not least because with the help of the matrix Y it is impossible, for example, to detect correctly the calculated components of the solutions of the systems considered in Examples 2.9.1–2.9.3. This, of course, is not by chance: the matrix Y is designed only to separate asymptotically stable linear combinations of the solution components.

Another serious limitation, when applying the matrix Y, is the requirement of constancy of the matrices involved in the system under investigation.

Thus, let us consider the difference problem

$$(A - \tau E)\frac{x_i - x_{i-1}}{\tau} = x_{i-1} + f_i, \quad i = 1, \dots, N; \qquad N\tau = \beta - \alpha, \qquad (2.9.52)$$

$$x(\alpha) = a,$$

where the matrix A is constant and is such that equation (2.7.6) has the solution Y with the properties indicated in the formulation of Lemma 2.7.1.

By applying Lemma 2.9.1 to the homogeneous equation

$$(A - \tau E)\frac{x_i - x_{i-1}}{\tau} = x_{i-1}, \qquad (2.9.53)$$

and putting $B = E$ and $M = Y$ in it, we obtain that if

$$-(A^{\mathscr{D}}A)^*(A^{\mathscr{D}}A) - \tau Y \leqslant 0, \qquad (2.9.54)$$

then, when $k \geqslant s$, for all solutions of equation (2.9.53) the estimate

$$(Yx_k, x_k) \leqslant (Yx_s, x_s) \qquad (2.9.55)$$

holds. However, since as a consequence of the permutability of the matrices A and $A^{\mathscr{D}}$ as well as of the equality $(A^{\mathscr{D}}A)^2 = A^{\mathscr{D}}A$, one can write $Y = (A^{\mathscr{D}}A)^*YA^{\mathscr{D}}A$ (see formula (2.7.10)). Then, by virtue of the positive definiteness of the matrix Y on the transform of the matrix $A^{\mathscr{D}}A$, the inequality (2.9.54) is automatically satisfied and, therefore, the inequality (2.9.55) is undoubtedly valid.

Since the matrix Y can be written as

$$Y = (A^{\mathscr{D}}A)^*ZA^{\mathscr{D}}A,$$

where the matrix Z is positively defined and is represented by the formula

$$Z = \int_0^\infty e^{A^*t} \cdot e^{At}\, dt,$$

then the inequality (2.9.55) can be rewritten as

$$(ZA^{\mathscr{D}}Ax_k, A^{\mathscr{D}}Ax_k) \leqslant (ZA^{\mathscr{D}}Ax_s, A^{\mathscr{D}}Ax_s).$$

But then

$$\|A^{\mathscr{D}}Ax_k\| \leqslant \sqrt{\mu^{-1}\lambda}\|A^{\mathscr{D}}Ax_s\|, \qquad (2.9.56)$$

where μ and λ are, respectively, the smallest and largest eigen-numbers of the matrix Z.

Returning to the inhomogeneous problem (2.9.52), we at first write its solution as

$$x_i = S^i x_0 + \tau \sum_{k=1}^i S^{i-k}(A - \tau E)^{-1}f_k. \qquad (2.9.57)$$

Then, by multiplying the equality (2.9.57) by the matrix $A^{\mathscr{D}}A$ and applying the inequality (2.9.56), we obtain:

$$\|A^{\mathscr{D}}Ax_i\| \leqslant \sqrt{\mu^{-1}\lambda}\|A^{\mathscr{D}}Ax_0\| + \tau\sqrt{\mu^{-1}\lambda}\sum_{k=1}^i \|A^{\mathscr{D}}A(A - \tau E)^{-1}f_k\|,$$

and for the deviation $v_i = x_i - x(i\tau)$ we obtain:

$$\|A^{\mathscr{D}}Av_i\| \leqslant \frac{\tau^2}{2}\sqrt{\mu^{-1}\lambda}\sum_{k=1}^i \|(A - \tau E)^{-1}A^{\mathscr{D}}A^2x''((k-1)\theta\tau)\|, \qquad (2.9.58)$$

from which, as before, it follows that $A^{\mathscr{D}}\cdot A\cdot v_i \to 0$ when $\tau \to 0$, i.e. the linear combinations $y_i = A^{\mathscr{D}}Ax_i$ converge.

Let us now assume that the index of the matrix A does not exceed unity. Then, by multiplying the equation from (2.9.52) by the matrix $E - A^{\mathscr{D}}A$, we obtain $(E - A^{\mathscr{D}}A)x_i = -(E - A^{\mathscr{D}}A)f_i$. But the same is obtained also by multiplying the differential equation $Ax' = x + f$ being approximated by the matrix $E - A^{\mathscr{D}}A$. Therefore $(E - A^{\mathscr{D}}A)v_i = 0$, i.e. $A^{\mathscr{D}}Av_i = v_i$ and, consequently, in view of the equality $A^{\mathscr{D}}A^2 = A$, the estimate (2.9.58) allows us to write the inequality

$$\|v_i\| \leqslant \frac{\tau^2}{2}\sqrt{\mu^{-1}\lambda}\sum_{k=1}^{i}\|(A - \tau E)^{-1}Ax''((k-1)\theta\tau)\|,$$

from which it follows that $v_i \to 0$ when $\tau \to 0$.

2.10. Summary of the obtained difference schemes and conditions for their applicability

The difference schemes constructed and substantiated in the preceding sections are of great practical value: they are rather simple and, under certain conditions, give the solution of singular problems:

$$A(t)x'(t) = B(t)x(t) + f(t), \quad \alpha \leqslant t \leqslant \beta, \quad x(\alpha) = a, \tag{2.10.1}$$

$$A(t)x'(t) = x(t) + f(t), \quad \alpha \leqslant t \leqslant \beta, \quad x(\alpha) = a. \tag{2.10.2}$$

Therefore, to ease the use of these schemes, we shall write them once more, indicating the conditions of their applicability, by making a general assumption that the solutions of problems (2.10.1) and (2.10.2) exist and are unique.

Let us, first of all, consider the difference schemes which approximate the problem (2.10.2): problem (2.10.1) can often be reduced to problem (2.10.2) through a simple substitution $x(t) = e^{ct}y(t)$, where $c = $ constant.

Scheme 1.

$$(A_i - \tau E)x_i = A_ix_{i-1} + \tau f_i, \quad i = 1,\ldots,N,$$
$$N\tau = \beta - \alpha, \quad x_0 = a. \tag{2.10.3}$$

This scheme is applicable when ind $A(t) \leqslant 1$ on the entire segment $[\alpha,\beta]$. The condition of its applicability (ind $A(t) \leqslant 1$) is satisfied, in particular, in the following cases:

(1) the equality $A = A(A^2)^- A^2$ or the equality $A = A^2(A^2)^- A$ holds (we shall sometimes omit to mention the dependence on t and i in the subsequent treatment);

(2) the matrix ΓB is nonsingular, where B and Γ are the components of the skeleton expansion $A = B\Gamma$;

(3) the inequality $A^* + A \leqslant 0$ is satisfied (in the sense of the inequalities between the relevant Hermitian forms); and

(4) the matrix $(A + \tau^2 E)^{-1}A$, when $\tau \to 0$, has a finite limit.

Scheme (2.10.3) is applicable also in the case when Ind $A(t) > 1$, but for all sufficiently small τ and all $i = 1, \ldots, N$, there exists a constant $K > 0$ (independent of i and τ), at which $\rho = 1 + K\tau$ for all i satisfies the inequality

$$A_i^* + A_i - \tau E + \frac{1 - \rho^2}{\tau \rho^2} A_i^* A_i \leqslant 0,$$

and, moreover, on the entire segment $[\alpha, \beta]$ the estimate

$$\tau \| (A_i - \tau E)^{-1} A_i x''((i - 1)\theta\tau) \| \leqslant \varepsilon,$$

is valid, where $x''((i - 1)\theta\tau)$ is the value of the second derivative of the accurate solution of the problem (2.10.2), and ε is the value of the required absolute accuracy of the solution (see the control example studied in the preceding section).

Note, furthermore, that if the matrix satisfying the inequality ind $A(t) \leqslant 1$ is constant, and the equation

$$A^* Y + YA = -(A^{\mathscr{D}} A)^* A^{\mathscr{D}} A$$

has a Hermitian positively defined (on the transform of the matrix $A^{\mathscr{D}} A$) solution, then the linear combinations $y = A^{\mathscr{D}} Ax$ are calculated to be asymptotically stable according to scheme (2.10.3).

Scheme 2.

$$(A_i^{k+1} - \tau A_i^k + \tau^2 E)x_i = A_i^{k+1}x_{i-1} - \tau^2 \sum_{s=0}^{k-1} A_i^s f_i^{(s)} + \tau A_i^k f_i, \quad i = 1, \ldots, N,$$

$$N\tau = \beta - \alpha, \qquad x_0 = a. \tag{2.10.4}$$

This scheme is applicable if ind $A(t) \leqslant k = \text{constant}$ on the entire segment $[\alpha, \beta]$ (it is clear that ind $A(t) \leqslant$ the order of the matrix), and, moreover, the matrix $(E - A^{\mathscr{D}} A)A$ is constant (the property Ω).

Note that the presence in scheme (2.10.4) of derivatives of the right-hand side of $f(t)$ makes the problem of computer analytic computations urgent.

Control examples:

(a) $\qquad A(t) = \begin{pmatrix} 0 & -t^{-1} & -t^{-1} - 1 & t^{-1} + 1 \\ 0 & 2 & 3 & -2 \\ 0 & -1 & -1 & 1 \\ 0 & 0 & 1 & 0 \end{pmatrix}, \qquad f(t) \equiv 0,$

$$0.5 \leqslant t \leqslant 1.5, \qquad \text{ind } A(t) = 3.$$

The accurate solution is $x(t) = (t^{-1}e^t; -e^t; e^t; e^t)^T$.

(b)
$$A(t) = \begin{pmatrix} 0 & 1 & 0 \\ 0 & 0 & 0 \\ 0 & e^{-t} & 1 \end{pmatrix}, \qquad f(t) = \begin{pmatrix} -e^{-t} - e^t \\ -e^{-t} \\ -e^{-2t} \end{pmatrix},$$

$$0 \leqslant t \leqslant 1, \qquad \text{ind } A(t) = 2.$$

The accurate solution is $x(t) = (e^t; e^{-t}; 0)^T$.

(c)
$$A(t) = \begin{pmatrix} 0 & 1 & e^t - 1 \\ 0 & 0 & 1 \\ 0 & 0 & 1 \end{pmatrix}, \qquad f(t) = \begin{pmatrix} e^{2t} - e^t \\ 0 \\ 0 \end{pmatrix},$$

$$0 \leqslant t \leqslant 6, \qquad \text{ind } A(t) = 2.$$

The accurate solution is $x(t) = (e^t; e^t; e^t)^T$.

Scheme 3.

$$(A_i^3 - \tau A_i^2 + \tau^2 E)x_i = A_i^3 x_{i-1} - \tau^4 C_i T_i A_i f_i' + \tau(A_i^2 - \tau E)f_i, \quad i = 1, \ldots, N,$$

$$N\tau = \beta - \alpha, \qquad x_0 = a, \tag{2.10.5}$$

where

$$C_i = \{A_i^* A_i + [E + \tau^2 T_i A_i']^* [E + \tau^2 T_i A_i']\}^{-1} [E + \tau^2 T_i A_i']^*,$$
$$T_i = (A_i^3 + \tau^2 E)^{-1}. \tag{2.10.6}$$

This scheme is applicable if

$$\text{ind } A(t) \leqslant 2, \qquad (E - A^{\mathscr{D}} A) AA' A^{\mathscr{D}} = 0,$$

and the matrix

$$A^* A + [E + (E - A^{\mathscr{D}} A) A']^* [E + (E - A^{\mathscr{D}} A) A'] \tag{2.10.7}$$

is reversible.

If the reversibility of the matrix (2.10.7) is due to the nonsingularity of the matrix $E + (E - A^{\mathscr{D}} A) A'$, then as the matrix (2.10.6) one can take the matrix $C_i = (E + \tau^2 T_i A_i')^{-1}$.

Control examples:

(a)
$$A = \begin{pmatrix} -t & -2t & 1 - 2t - t^2 & -t \\ 0 & 0 & 0 & 0 \\ 0 & 0 & 0 & 0 \\ 1+t & 2(1+t) & (1+t)^2 & 1+t \end{pmatrix}, \qquad 0 \leqslant t \leqslant 3,$$

$$f(t) = \begin{pmatrix} -(3t + t^2)e^t \\ 0 \\ -e^t \\ 2 + 3t + t^2 \end{pmatrix}.$$

The accurate solution is $x(t) = (e^t; 0; e^t; 0)^T$.

(b) $A = \begin{pmatrix} 0 & 1+t & -(1+t) \\ 0 & 0 & 2 \\ 0 & 0 & 1 \end{pmatrix}$, $f(t) = \begin{pmatrix} -(2+t)e^t \\ 2e^t \\ 0 \end{pmatrix}$, $0 \le t \le 3$.

The accurate solution is $x(t) = (e^t; 0; e^t)^T$.

Scheme 4.

This scheme coincides in its outer appearance with the scheme (2.10.5), but the matrix C_i is calculated by the formula

$$C_i = [A_i^* A_i + (E + A_i')^*(E + A_i')]^{-1}(E + A_i')^*.$$

The applicability conditions are ind $A(t) \le 2$, the matrix $A^{\mathscr{D}}A$ is constant, and the matrix $A^*A + (E + A')^*(E + A')$ is reversible.

If the matrix $E + A'$ is reversible, then the matrix $C_i = (E + A_i')^{-1}$ can be taken as the matrix C_i.

Control example:

$$A(t) = \begin{pmatrix} 1+t & 1+t & -(1+t) \\ -1 & 0 & 1 \\ t & 1+t & -t \end{pmatrix}; \quad f(t) = \begin{pmatrix} te^t \\ -e^t \\ te^t \end{pmatrix}, \quad 0 \le t \le 1.$$

The accurate solution is $x(t) = (e^t; e^t; e^t)^T$.

Scheme 5.

$$[A_i^3 - \tau A_i^2 + \tau^2 E - \tau^3 A_i^2 A_i' M_i]z_i = A_i^3 z_{i-1} + [\tau A_i^2 - \tau^2 E + \tau^3 A_i^2 A_i' M_i]f_i$$
$$- \tau^3 A_i^2 A_i' C_i T_i A_i f_i', \quad i = 1, \ldots, N;$$

$$N\tau = \beta - \alpha, \qquad z_0 = T_0(A_0^3 a - \tau^2 f_0), \tag{2.10.8}$$
$$x_i = (E + \tau^2 M_i)z_i + \tau^2 M_i f_i - \tau^2 C_i T_i A_i f_i',$$

where the matrices C_i and T_i are calculated by formulae (2.10.6), and

$$M_i = C_i T_i A_i A_i' T_i A_i^2.$$

This scheme is applicable when ind $A(t) \le 2$ and the matrix (2.10.7) is nonsingular.

If the nonsingularity of the matrix (2.10.7) is due to the nonsingularity of the matrix $(E + (E - A^{\mathscr{D}}A)A')$, then the matrix $C_i = (E + \tau^2 T_i A_i')^{-1}$ can be taken as the matrix C_i in (2.10.8).

Control examples:

(a) $A(t) = \begin{pmatrix} t & 1 & 1+t \\ -t^2 & -t & 0 \\ 0 & 0 & 1+t \end{pmatrix}$, $f(t) = \begin{pmatrix} (2t+1)e^t \\ -(1+t+t^2)e^t \\ te^t \end{pmatrix}$, $0 \le t \le 1$.

The accurate solution is $x(t) = (e^t; e^t; e^t)^T$.

(b) $A(t) = \begin{pmatrix} t & 1 & 1+t \\ -t^2 & -t & 0 \\ 0 & 0 & 1 \end{pmatrix}$, $f(t) = \begin{pmatrix} 2te^t \\ -t^2e^t \\ 0 \end{pmatrix}$, $0 \leqslant t \leqslant 1$.

The accurate solution is $x(t) = (e^t; 0; e^t)^T$.

We now turn to the difference schemes which approximate the problem (2.10.1).

Scheme 6.

$$x_i = (A_i^* A_i - \tau A_i^* B_i)^{-1}(A_i^* A_i x_{i-1} + \tau A^* f_i), \quad i = 1,\ldots,N,$$
$$N\tau = \beta - \alpha, \qquad x_0 = a.$$

This scheme is applicable if for all $t \in [\alpha, \beta]$ the columns of the $(m \times n)$ matrix $A(t)$ are linearly independent.

Control example:

$$A(t) = \begin{pmatrix} 1 & t \\ t & 1+t^2 \\ t & 1+t^2 \end{pmatrix}, \quad B(t) = \begin{pmatrix} 0 & 0 \\ 1 & t \\ t & t^2 \end{pmatrix}, \quad f(t) = \begin{pmatrix} e^t \\ (t-1)e^t \\ 0 \end{pmatrix}, \quad 0 \leqslant t \leqslant 1.$$

The accurate solution is $x(t) = (e^t; 0)^T$.

Difference schemes for solving the problem (2.10.1) can be constructed under the assumption that the columns of the $(m \times n) - B(t)$ matrix are linearly independent (this can sometimes be achieved through the substitution $x(t) = e^{ct}y(t)$, $c = $ constant, and problem (2.10.1) is equivalent to the problem

$$B^+(t)A(t)x'(t) = x(t) + B^+(t)f(t), \quad \alpha \leqslant t \leqslant \beta, \quad x(\alpha) = a. \tag{2.10.9}$$

In particular, this equivalence occurs when the pair of matrices (B, A) is quite perfect, i.e. $(E - BB^+)A = 0$.

In order to solve problem (2.10.9), one can try to apply one of the difference schemes 1–5 considered above (if, of course, the matrix B^+A satisfies the relevant requirements), and in order to get rid of pseudo-inverse matrices it is natural to make use of the formula $B^+ = (B^*B)^{-1}B^*$. In this case, for example, the difference scheme (2.10.3) receives the form

$$(B_i^* A_i - \tau B_i^* B_i)x_i = B_i^* A_i x_{i-1} + \tau B_i^* f_i, \quad i = 1,\ldots,N;$$
$$N\tau = \beta - \alpha, \qquad x_0 = a.$$

Control example:

$$A(t) = \begin{pmatrix} 0 & 0 \\ -\sin t \cos t & \cos^2 t \\ \sin^2 t & -\sin t \cos t \end{pmatrix},$$

$$B(t) = \begin{pmatrix} \cos t & \sin t \\ -\sin t \cos t & \cos^2 t \\ \sin^2 t & -\sin t \cos t \end{pmatrix},$$

$$f(t)(-e^t \sin t - e^t \cos t; 0; 0)^T, \quad 0 \leqslant t \leqslant 12, \quad B^*B = E.$$

The accurate solution is $x(t) = (e^t; e^t)^T$.

We give also an example of the matrix $A(t)$ at which the system (2.10.2) does not satisfy the applicability conditions of the schemes mentioned above:

$$A(t) = \begin{pmatrix} 0 & e^{-t} & -e^{-2t} & 0 \\ 0 & 0 & 1 & 0 \\ 0 & 0 & 0 & 0 \\ 0 & 0 & -e^{-t} & 0 \end{pmatrix}, \qquad (2.10.10)$$

$$\text{ind } A(t) = 3, \qquad (E - A^{\mathscr{D}} A)A \neq \text{constant} \qquad A^{\mathscr{D}} A \neq \text{constant}.$$

However, the solution of the system (2.10.2) with the matrix (2.10.10) is readily obtained using the elimination method. For systems with constant matrices, this method is studied in detail in [23, p. 23] as well as in [24], where a study is made of the relationship between the number of elimination steps and the structure of the pair of matrices of the system. For three-step systems, the elimination method was realized in the program [25].

To conclude, we must note that the difference schemes 1–6 proposed here are also applicable in the case of a variable step τ, which cannot be said, for example, with regard to the implicit Euler scheme (see [26]).

3 SOME METHODS OF FINDING GENERAL SOLUTIONS OF BOUNDARY-VALUE PROBLEMS ASSOCIATED WITH SINGULAR SYSTEMS OF LINEAR ORDINARY DIFFERENTIAL EQUATIONS

This chapter considers a number of problems amenable to solution with the help of generalized inverse matrices. The formulae obtained here for general solutions and the results of the preceding chapters can provide the basis when constructing the new difference schemes.

For the subsequent discussion, we shall agree on a number of symbols. A set of summable, continuous, absolutely continuous functions on the segment $[\alpha, \beta]$ will be denoted, respectively, by Σ, $\mathfrak{R}$, and $\mathfrak{A}$. If $M(t)$ is a matrix, in particular, a vector, then the representations $M(t) \in \Sigma$, $M(t) \in \mathfrak{R}$, and $M(t) \in \mathfrak{A}$ will mean that all elements of the matrix $M(t)$ belong, respectively, to Σ, $\mathfrak{R}$, and $\mathfrak{A}$.

3.1. Conditions imposed on the solutions

Additional conditions, which must be satisfied by the desired solutions of singular systems, will be written in this chapter with the help of Stiltjes integrals, namely

$$\int_\alpha^\beta [d\sigma(s)]C(s)x(s) = a, \tag{3.1.1}$$

where a is a given vector; $C(s)$ is a given matrix with continuous elements on $[\alpha, \beta]$; $\sigma(s)$ is a given matrix the elements of which are real on the $[\alpha, \beta]$ function with bounded full variation (all the other quantities in (3.1.1) and in the singular system can be complex); and $x(s)$ is the desired continuous vector. The meaning of the condition (3.1.1) implies the following: if $C(s)$ is a $(q \times n)$ matrix with elements $c_{jk}(s)$, $\sigma(s)$ is a $(p \times q)$ matrix with elements $\sigma_{ij}(s)$, a is a vector of dimension p with components a_i, and $x(s)$ is a vector of dimension n with components $x_k(s)$, then the equality (3.1.1) is a system of the following scalar equalities:

$$\sum_{j=1}^q \int_\alpha^\beta [d\sigma_{ij}(s)] \sum_{k=1}^n c_{jk}(s)x_k(s) = a_i, \quad i = 1, \dots, p. \tag{3.1.2}$$

The representation of conditions with the help of the system of equalities (3.1.2) covers, in particular, a multi-point boundary-value problem and the Cauchy problem. In order to specify, for example, a multi-point boundary-value problem, it is sufficient to take as the matrix $\sigma(s)$ the square diagonal matrix the elements of which are stepped functions of the form

$$\psi(s) = \begin{cases} \varepsilon, & s = s_0 = \alpha, \\ \eta_0, & \alpha < s \leqslant s_1, \\ \eta_1, & s_1 < s \leqslant s_2, \\ \cdots & \\ \eta_{N-1}, & s_{N-1} < s < s_N = \beta, \\ \eta_N, & s = s_N = \beta. \end{cases} \tag{3.1.3}$$

In this connection, it must be recalled that if the function $\psi(s)$ is a function of the form (3.1.3), and $\varphi(s)$ is a continuous function, then

$$\int_\alpha^\beta [d\psi(s)]\varphi(s) = (\eta_0 - \varepsilon)\varphi(\alpha) + (\eta_1 - \eta_0)\varphi(s_1) + \cdots + (\eta_N - \eta_{N-1})\varphi(\beta),$$

i.e. in this case the Stiltjes integral equals the sum of $N + 1$ summands, each of which is the product of the value of the discontinuity of the function $\psi(s)$ by the value of the function $\varphi(s)$ at the point s_i ($i = 0, 1, \dots, N$) corresponding to this jump. Thus, in the case of the Cauchy problem, as the matrix $\sigma(s)$ in (3.1.1) one should take a scalar diagonal matrix, the diagonal elements of which equal the function

$$\psi(s) = \begin{cases} 0, & s = \alpha, \\ 1, & \alpha < s \leqslant \beta, \end{cases}$$

and, at the same time, put $C(s) \equiv E$. With such a choice of the matrices $\sigma(s)$ and $C(s)$, by virtue of formula (3.1.2), we obtain:

$$\int_\alpha^\beta [d\sigma(s)] x(s) = x(\alpha),$$

and, therefore, the Cauchy condition $x(\alpha) = a$ can be written as

$$\int_\alpha^\beta [d\sigma(s)] x(s) = a.$$

Note, furthermore, that the representation of the boundary-value and other conditions in the form of the Stiltjes integral is not only convenient but it also allows us to calculate all problems which permit the representation of the conditions in the form (3.1.1) related in the sense that, from the point of view of the methods to solve them, they are nondistinguishable.

3.2. The main relationship between the values of the solution of a system of linear ordinary differential equations

Let $x(t) \in \mathfrak{A}$ be the solution of the system

$$x'(t) = B(t)x(t) + f(t), \quad \alpha \leqslant t \leqslant \beta, \qquad (3.2.1)$$

in which $b \in \Sigma$ and $f \in \Sigma$. We denote by $X(t)$ the matrix of fundamental solutions of a corresponding homogeneous system. As is known, as $X(t)$ one can take a matrixant, i.e. the solution of the matrix Cauchy problem (see [1, p. 429]):

$$X'(t) = B(t)X(t), \quad \alpha \leqslant t \leqslant \beta,$$
$$X(\alpha) = E. \qquad (3.2.2)$$

The matrixant $X(t)$ is defined by the series

$$X(t) = E + \int_\alpha^t B(t_1)\,dt_1 + \int_\alpha^t B(t_2) \int_\alpha^{t_2} B(t_1)\,dt_1\,dt_2 + \cdots \qquad (3.2.3)$$

and, for all $t \in [\alpha, \beta]$, is a nonsingular matrix.

Using (3.2.2) and (3.2.3) it is easy to show that all elements of the matrixant $X(t)$ as well as the elements of the matrix $X^{-1}(t)$ inverse to it, are absolutely continuous functions on $[\alpha, \beta]$ such that

$$X(t) \in \mathfrak{A}, \qquad X^{-1}(t) = \mathfrak{A}.$$

For the solution $x \in \mathfrak{A}$ of the system (3.2.1), the equality

$$[X(\tau)X^{-1}(\tau)x(\tau)]' = B(\tau)X(\tau)X^{-1}(\tau)x(\tau) + f(\tau), \quad \alpha \leqslant \tau \leqslant \beta$$

is obviously valid (almost everywhere on $[\alpha, \beta]$) or, after taking the derivative, the equality

$$X'(\tau)X^{-1}(\tau)x(\tau) + X(\tau)[X^{-1}(\tau)x(\tau)]' = B(\tau)X(\tau)X^{-1}(\tau)x(\tau) + f(\tau). \qquad (3.2.4)$$

If now the identity (3.2.2) is taken into consideration, then with the help of the equality (3.2.4), we obtain:

$$[X^{-1}(\tau)x(\tau)]' = X^{-1}(\tau)f(\tau). \qquad (3.2.5)$$

By integrating the equality (3.2.5) in the limits from s to t $(s, t \in [\alpha, \beta])$ and multiplying the obtained equality by the matrix $X(t)$, we arrive at the relationship

$$x(t) = X(t)X^{-1}(s)x(s) + \int_s^t X(t)X^{-1}(\tau)f(\tau)\,d\tau. \qquad (3.2.6)$$

Thus, for the values of $x(t)$ and $x(s)$ of any solution of $x(t) \in \mathfrak{A}$ of the system (3.2.1), where s and t are arbitrary points of the segment $[\alpha, \beta]$, the relationship (3.2.6) is valid.

Let us now apply the obtained relationship (3.2.6) to solve the problem

$$x'(t) = B(t)x(t) + f(t), \quad \alpha \leqslant t \leqslant \beta, \qquad (3.2.7)$$

$$\int_\alpha^\beta [d\sigma(s)]C(s)x(s) = a, \qquad (3.2.8)$$

where $B \in \Sigma$ and $f \in \Sigma$. The problem is that of finding a vector $x \in \mathfrak{A}$ that satisfies equation (3.2.7) (almost everywhere on $[\alpha, \beta]$) and the condition (3.2.8).

In order to solve this problem, we substitute the relationship (3.2.6) (by preliminarily replacing s and t) into the condition (3.2.8). As a result, with respect to $x(t)$ we obtain the algebraic system

$$\Gamma X^{-1}(t)x(t) = a + \varphi(t), \qquad (3.2.9)$$

in which

$$\Gamma = \int_\alpha^\beta [d\sigma(s)]C(s)X(s), \qquad (3.2.10)$$

$$\varphi(t) = \int_\alpha^\beta [d\sigma(s)]C(s)X(s) \int_s^t X^{-1}(\tau)f(\tau)\,d\tau. \qquad (3.2.11)$$

Thus, we have proved

Lemma 3.2.1 Any solution $x \in \mathfrak{A}$ of the problem (3.2.7), (3.2.8) satisfies the algebraic system (3.2.9), in which Γ and $\varphi(t)$ are calculated by formulae (3.2.10) and (3.2.11), and $X(t)$ is the solution of the problem (3.2.2).

Let us now fix the arbitrary point $t_0 \in [\alpha, \beta]$ and solve equation (3.2.9) for $t = t_0$. The vector $x(t_0)$ obtained is then substituted into the relationship (3.2.6) by preliminarily replacing s for t_0 in it. As a result, we obtain:

$$x(t) = X(t)X^{-1}(t_0)x(t_0) + \int_{t_0}^{t} X(t)X^{-1}(\tau)f(\tau)\,d\tau. \tag{3.2.12}$$

By means of a direct substitution (in view of (3.2.9)–(3.2.11)) it is easy to make sure that the vector $x(t)$ calculated by formula (3.2.12) is the solution of the problem (3.2.7), (3.2.8) and, consequently, according to Lemma 3.2.1, for all $t \in [\alpha, \beta]$, satisfies the algebraic system (3.2.9). Thus, if $x(t_0)$ runs through all solutions of the algebraic system (3.2.9) for $t = t_0$, then by formula (3.2.12) we obtain all solutions of the problem (3.2.7), (3.2.8).

As is already known (see Corollary 1.3.1), all solutions of the algebraic system (3.2.9) for $t = t_0$ can be obtained by the formula

$$x(t_0) = X(t_0)\Gamma^{-}[a + \varphi(t_0)] + X(t_0)\gamma, \tag{3.2.13}$$

where γ is an arbitrary solution of the system

$$\Gamma\gamma = 0, \tag{3.2.14}$$

and the compatibility condition of the system (3.2.9) is the equality

$$(E - \Gamma\Gamma^{-})[a + \varphi(t_0)] = 0. \tag{3.2.15}$$

Thus, the following theorems are proved.

Theorem 3.2.1   The problem (3.2.7), (3.2.8) has a solution if and only if at least at one point $t_0 \in [\alpha, \beta]$ the equality (3.2.15) is satisfied.

Remark 3.2.1   If the compatibility condition (3.2.15) is satisfied at some point $t_0 \in [\alpha, \beta]$, then it is satisfied also at all points of the segment $[\alpha, \beta]$. This follows from the existence of the solution (3.2.12), (3.2.13) and from Lemma 3.2.1, according to which the solution (3.2.12), (3.2.13) at all points of the segment $[\alpha, \beta]$ satisfies the system (3.2.9) and, consequently, at all $t \in [\alpha, \beta]$ the system (3.2.9) is compatible. Thus, as the point t_0 in (3.2.12), (3.2.13) we can take any point $t_0 \in [\alpha\beta]$. This remark should be borne in mind also in the subsequent treatment.

Theorem 3.2.2   The general solution of the problem (3.2.7), (3.2.8) (if it exists) is represented as

$$x(t) = X(t)\Gamma^{-1}[a + \varphi(t_0)] + X(t)\gamma + \int_{t_0}^{t} X(t)X^{-1}(\tau)f(\tau)\,d\tau,$$

where t_0 is an arbitrary fixed point of the segment $[\alpha, \beta]$, the matrix $X(t)$ is the solution of the problem (3.2.2), γ is an arbitrary solution of the system (3.2.14), and Γ and $\varphi(t_0)$ are calculated by formulae (3.2.10) and (3.2.11).

Theorem 3.2.3 The problem (3.2.7), (3.2.8) has only one solution if and only if $\gamma = 0$ is the unique solution of the system (3.2.14).

Remark 3.2.2 If the solution of the problem (3.2.7), (3.2.8) is unique, then it can be obtained by the formula

$$x(t) = X(t)\Gamma^+[a + \varphi(t)]$$

(see (3.2.13)) or (approximately) by the formula

$$x_\varepsilon(t) = X(t)(\Gamma^*\Gamma + \varepsilon E)^{-1}\Gamma^*(a + \varphi(t)).$$

3.3. The problem with a perfect group of three matrices

To begin with, we shall formulate the problem.

Problem 3.3.1 Let the group of three matrices $(A(t), B(t), C(t))$ be perfect (see Definition 1.7.1) and $f(t) \in \Sigma$.
Find the vector $x(t) \in \mathfrak{A}$ that satisfies the system

$$A(t)x'(t) = B(t)x(t) + f(t), \quad \alpha \leqslant t \leqslant \beta \tag{3.3.1}$$

(almost everywhere on $[\alpha, \beta]$) and the condition

$$\int_\alpha^\beta [d\sigma(s)] C(s)x(s) = a. \tag{3.3.2}$$

Subsequently, for brevity of the representations, we often omit that the matrices and vectors depend on t. The case of constancy of the matrices and vectors will be specially qualified.
The following lemma is valid.

Lemma 3.3.1 If $x \in \mathfrak{A}$ is the solution of Problem 3.3.1, then the pair of vectors (x, u), where

$$u = (E - A^- A)x', \tag{3.3.3}$$

satisfies the system

$$x'(t) = A^-(t)B(t)x(t) + A^-(t)f(t) + u(t), \tag{3.3.4}$$

$$[E - A(t)A^-(t)]B(t)x(t) = -[E - A(t)A^-(t)]f(t), \tag{3.3.5}$$

$$A(t)u(t) = 0, \quad \alpha \leqslant t \leqslant \beta \tag{3.3.6}$$

(almost everywhere on $[\alpha, \beta]$) and condition (3.3.2).
Conversely, if the pair $(x \in \mathfrak{A}, u \in \Sigma)$ satisfies the system (3.3.4)–(3.3.6) and condition (3.3.2), then the vector x from it is the solution of Problem 3.3.1, and the vector u is expressed in terms of the vector x by formula (3.3.3).

Proof That the pair (x, u), in which x is the solution of Problem 3.3.1 and u is expressed in terms of x by formula (3.3.3), is the solution of equations (3.3.4) and (3.3.6), is established through a simple substitution, in view of the equality (3.3.1) and the determining property $A(E - A^- A) = 0$ of the semi-inverse matrices. As far as equation (3.3.5) is concerned, it is obtained after multiplying the equality (3.3.1) by the matrix $E - AA^-$.

In order to prove the inverse assertion, we first multiply the equality (3.3.4) by the matrix $E - A^- A$. Then, by virtue of (3.3.5) and (3.3.6), we obtain:

$$(E - A^- A)x' = (E - A^- A)[A^- Bx + A^- f + u]$$
$$= A^-(E - AA^-)[Bx + f] + u = u,$$

i.e. (3.3.3) holds true.

We now multiply the equality (3.3.4) by the matrix A. As a result (and as a consequence of the assumed validity of the equalities (3.3.5), (3.3.6)), we obtain the equality (3.3.1). The lemma is thereby proved.

Thus, the solution of Problem 3.3.1 is reduced to solving the problem (3.3.4)–(3.3.6), (3.3.2).

In order to clarify the problem (3.3.4)–(3.3.6), (3.3.2), we write the relationship between the values of the solution of equation (3.3.4) as

$$x(t) = X(t)X^{-1}(s)x(s) + \int_s^t X(t)X^{-1}(\tau)[A^-(\tau)f(\tau) + u(\tau)]\,d\tau, \qquad (3.3.7)$$

where the matrix $X(t)$ is the solution of the problem

$$X'(t) = A^-(t)B(t)X(t), \qquad X(\alpha) = E. \qquad (3.3.8)$$

By permuting s and t in the equality (3.3.7) as well as replacing t by s in (3.3.5), we substitute (3.3.7) into the equalities (3.3.5) and (3.3.2). As a result, in view of the equality $u(\tau) = [E - A^-(\tau)A(\tau)]u(\tau)$ and the perfectness of the group of three matrices (A, B, C) we obtain:

$$G(s)X^{-1}(t)x(t) = G(s)\Phi(s, t) - [E - A(s)A^-(s)]\cdot f(s), \qquad (3.3.9)$$

$$\Gamma X^{-1}(t)x(t)x(t) = a + \varphi(t), \qquad \alpha \leqslant t \leqslant \beta, \quad \alpha \leqslant s \leqslant \beta, \qquad (3.3.10)$$

where

$$G(s) = [E - A(s)A^-(s)]B(s)X(s),$$

$$\Gamma = \int_\alpha^\beta [d\sigma(s)]C(s)X(s),$$

$$\Phi(s, t) = \int_s^t X^{-1}(\tau)A^-(\tau)f(\tau)\,d\tau, \qquad (3.3.11)$$

$$\varphi(t) = \int_\alpha^\beta [d\sigma(s)]C(s)X(s)\Phi(s, t).$$

Lemma 3.3.2 If the group of three matrices (A, B, C) is perfect, then any solution of the problem (3.3.4)–(3.3.6), (3.3.2) satisfies the system (3.3.10), (3.3.9), (3.3.6).

By fixing now the arbitrary point $t_0 \in [\alpha; \beta]$, we find the solution of the system (3.3.9), (3.3.10) independent of s for $t = t_0$. We substitute the vector $x(t_0)$ obtained into the relationship (3.3.7) by preliminarily replacing s by t_0 in it. As a result, we obtain:

$$x(t) = X(t)X^{-1}(t_0)x(t_0) + \int_{t_0}^{t} X(t)X^{-1}(\tau)[A^-(\tau)f(\tau) + u(\tau)]\,d\tau, \quad (3.3.12)$$

where $u(t)$ is the solution of equation (3.3.6).

Through a direct substitution (in view of the perfectness of the group of three matrices (A, B, C) and the equalities (3.3.9), (3.3.10) which are satisfied by the vector $x(t_0)$), it is easy to see that the vector $x(t)$ calculated by formula (3.3.12) is the solution of the problem (3.3.4)–(3.3.6), (3.3.2) and, consequently, according to Lemma 3.3.2 the pair of vectors (x, u) satisfies the system (3.3.9), (3.3.10), (3.3.6). Thus, if $x(t_0)$ runs through all s-independent solutions of the system (3.3.9), (3.3.10) when $t = t_0$, and $u(t)$ runs through all solutions of equation (3.3.6), then the vector (3.3.12) will run through all solutions of the problem (3.3.4)–(3.3.6), (3.3.2).

Let us now apply Lemmas 1.3.3–1.3.5 to the system (3.3.9), (3.3.10) when $t = t_0$. As a result we obtain that if the system is solvable, i.e. if there exists a vector $x(t_0)$ independent of s and satisfying the system (3.3.9), (3.3.10), then its general (constant) solution $x(t_0)$ is represented by the formula

$$x(t_0) = X(t_0) \cdot \Gamma_1^- \cdot \Gamma_2(t_0) + X(t_0)\gamma, \quad (3.3.13)$$

where

$$\Gamma_1 = \int_{\alpha}^{\beta} [G^*(s)G(s) + \Gamma^*\Gamma]\,ds,$$

$$\Gamma_2(t_0) = \int_{\alpha}^{\beta} [G^*(s)G(s)\Phi(s, t_0) + \Gamma^*a + \Gamma^*\varphi(t_0) - G^*(s)[E - A(s)A^-(s)]f(s)]\,ds,$$

$$(3.3.14)$$

in which γ is an arbitrary (constant) solution of the system

$$\Gamma_1\gamma = 0 \quad (3.3.15)$$

(for the other notations see in (3.3.8) and (3.3.11)).

The compatibility condition of the system (3.3.9), (3.3.10) when $t = t_0$, according to Lemma 1.3.4, is the validity of the equalities

$$G(s)\Gamma_1^-\Gamma_2(t_0) = G(s)\Phi(s, t_0) - [E - A(s)A^-(s)]f(s),$$

$$\Gamma\Gamma_1^-\Gamma_2(t_0) = a + \varphi(t_0), \quad \alpha \leqslant s \leqslant \beta. \quad (3.3.16)$$

Thus, the following theorems can be considered proved.

Theorem 3.3.1   Problem 3.3.1 is solvable if and only if at some point $t_0 \in [\alpha, \beta]$ the system of inequalities (3.3.16) is satisfied.

Theorem 3.3.2   The general solution of Problem 3.3.1 (if it exists) is represented as

$$x(t) = X(t)X^{-1}(t_0)x(t_0) + \int_{t_0}^{t} X(t)X^{-1}(\tau)[A^-(\tau)f(\tau) + u(\tau)]\,d\tau,$$

$$x(t_0) = X(t_0)\Gamma_1^- \Gamma_2(t_0) + X(t_0)\gamma,$$

$$(3.3.17)$$

where t_0 is an arbitrary fixed point of the segment $[\alpha, \beta]$, the matrix $X(t)$ is the solution of the problem (3.3.8), γ is an arbitrary solution of the system (3.3.15), and Γ_1 and $\Gamma_2(t_0)$ are calculated by formulae (3.3.14) and (3.3.11).

Theorem 3.3.3   Problem 3.3.1 has only one solution if and only if $\gamma = 0$ and $u(\tau) = 0$ are unique solutions, respectively, of the systems (3.3.14) and (3.3.6) (this requirement means a linear independence of the matrix columns of the systems (3.3.14), (3.3.16).

Obviously, using formula (3.3.17) and the pseudo-inverse matrices with their approximations of the form (1.2.20) for the solution of Problem 3.3.1, one can obtain difference schemes. If necessary, the reader can do this easily.

Note further that if the solution of Problem 3.3.1 is unique, then it can be obtained by the formula

$$x(t) = X(t)\Gamma_1^+ \Gamma_2(t)$$

(see (3.3.13) and Remark 3.2.2)), from which in the case of the Cauchy problem with an initial given $x(\alpha) = a$ it follows, for example, the matching condition of the vector a with the right-hand side of $f(t)$, namely

$$\int_{\alpha}^{\beta} G^*(s)G(s)\,ds \cdot a$$

$$= \int_{\alpha}^{\beta} \left\{ G^*(s)G(s) \int_{s}^{\alpha} X^{-1}(\tau)A^-(\tau)f(\tau)\,d\tau - G^*(s)[E - A(s)A^-(s)]f(s) \right\} ds.$$

3.4.   Systems of the general form. Application of a resolving pair of matrices

Let us consider the system

$$A(t)x'(t) = B(t)x(t) + f(t), \quad \alpha \leqslant t \leqslant \beta, \tag{3.4.1}$$

by assuming that $A' \in \mathfrak{R}$, $B' \in \mathfrak{R}$, $f \in \Sigma$ and that there exists a resolving pair of matrices (A^B, Y) corresponding to the pair of matrices (A, B) such that $Y' \in \mathfrak{R}$,

$[(YA)^{\mathscr{D}}]'\in\mathfrak{R}$ and, consequently, $(A^B) = [(YA)^{\mathscr{D}}Y]'\in\mathfrak{R}$. Under these assumptions, by the solution of the system (3.4.1), we naturally understand an absolutely continuous vector $x(t)$, satisfying (almost everywhere on $[\alpha, \beta]$) the system (3.4.1). Suppose, furthermore, that $\mathrm{ind}\,[Y(t)A(t)] \leqslant k = \mathrm{constant}$. It is clear that $\mathrm{ind}(YA) \leqslant A$ is of the order of the matrix YA.

The following lemma is valid.

Lemma 3.4.1 The system (3.4.1) is equivalent to the system

$$A^B A x' = A^B B x + A^B f, \tag{3.4.2}$$

$$(E - A^B A)Y A x' = (E - A^B A)Y B x + (E - A^B A)Y f, \tag{3.4.3}$$

$$(E - BY)A x' = (E - BY)f \tag{3.4.4}$$

(for brevity, we omit indicating the dependence on $t\in[\alpha, \beta]$).

Proof The fact that the system (3.4.2)–(3.4.4) follows from equation (3.4.1) is obvious: the equalities (3.4.2)–(3.4.4) are obtained after multiplying equation (3.4.1) by the matrices A^B, $(E - A^B A)Y$, and $(E - BY)A$, respectively, in view of the fact that the matrix Y is a semi-inverse matrix of the matrix B. In order to prove that (3.4.1) follows from (3.4.2)–(3.4.4), we multiply equation (3.4.2) by the matrix YA. Then, taking into account the equality $YAA^B = A^B AY$ (see (1.10.10)), we obtain:

$$A^B A Y A x' = A^B A Y B x + A^B A Y f. \tag{3.4.5}$$

By adding now (3.4.5) to (3.4.3) and multiplying the obtained result by the matrix B, we arrive at the equality

$$BY A x' = x + BY f. \tag{3.4.6}$$

Finally, we note that the sum of the equalities (3.4.6) and (3.4.4) is the equality (3.4.1). The lemma is proved.

Now, for brevity of notation, we introduce the designations

$$\eta = A^B A x, \qquad \zeta = (E - A^B A)x, \qquad u = (E - YB)x. \tag{3.4.7}$$

Then, in view of the equalities (3.4.7) and

$$A^B B = A^B B A^B A, \qquad A^B A Y B = A^B A,$$

$$A^B A Y A = Y A A^B A, \qquad A^B A (A^B A)' A^B A = 0, \qquad BY A A^B = A A^B$$

(see, respectively, (1.10.21), (1.10.15), (1.10.24), (1.10.30), and (1.10.16)), the system (3.4.2)–(3.4.4) is written as

$$\eta' = [A^B B + (A^B A)']\eta + (A^B A)'\zeta + A^B f,$$

$$(E - A^B A)Y A \zeta' = \zeta - (E - A^B A)Y A (A^B A)'\eta - u + (E - A^B A)Y f,$$

$$(E - BY)A \zeta' = [(E - BY)A]'\eta + (E - BY)f.$$

The next step involves replacing the matrix $(A^B A)'$ with its representation (1.10.29). As a result of this replacement, and after some transformations associated with differentiation and use of the equality $(E - A^B A)(YA)^k = 0$ (see (1.10.25)), we obtain the system

$$\eta' = [A^B B + (E - A^B A)YA((YA)^{k-1})'((YA)^{\mathscr{D}})^k$$
$$+ (E - A^B A)(YA)'(YA)^{\mathscr{D}}]\eta + [((YA)^{\mathscr{D}})^k((YA)^{k-1})' YA(E - A^B A)$$
$$+ (YA)^{\mathscr{D}}(YA)'(E - A^B A)]\zeta + A^B f, \qquad (3.4.8)$$

$$(E - A^B A)YA\zeta' = \zeta - (E - A^B A)YA((YA)^{k-1})'((YA)^{\mathscr{D}})^{k-1}\eta$$
$$- u + (E - A^B A)Yf, \qquad (3.4.9)$$

$$(E - BY)A\zeta' = [(E - BY)A]'\eta + (E - BY)f. \qquad (3.4.10)$$

Proceeding in the same way, and using the properties of the resolving pair of matrices and the equality (1.10.29), one can go from the system (3.4.8)–(3.4.10) to the system (3.4.2)–(3.4.4) and, consequently, the systems (3.4.8)–(3.4.10) and (3.4.2)–(3.4.4) are equivalent. Thus, on the basis of Lemma 3.4.1 the solution of equation (3.4.1) is brought to the solution of the system (3.4.8)–(3.4.10).

Note that the factor $E - A^B A$ in the second square brackets of the right-hand side of equation (3.4.8) is actually superfluous because $A^B Au = A^B A(E - YB)x = 0$ and using equation (3.4.9) it is easy to show that $A^B A\zeta = 0$. This factor is retained here in order to reveal quickly 'good' classes of systems.

3.5. The property Ω. The main relationship between values of the solution of systems with the property Ω

Definition 3.5.1 It will be said that the pair of matrices $(A(t), B(t))$ defined on the segment $[\alpha, \beta]$ has the property Ω on this segment if: (1) at each fixed $t \in [\alpha, \beta]$ the pair of matrices $(B(t), A(t))$ is imperfect; and (2) for the pair of matrices $(A(t), B(t))$ there exists such a resolving pair of matrices $(A^B(t), Y(t))$ that the matrices

$$(E - A^B A)YA, \qquad (E - BY)A \qquad (3.5.1)$$

are constant and, moreover,

$$A' \in \mathfrak{N}, \qquad B' \in \mathfrak{N}, \qquad Y' \in \mathfrak{N}, \qquad [(YA)^{\mathscr{D}}]' \in \mathfrak{N}.$$

Note that requirement (1) is not a limiting one, because on the basis of Theorem 1.6.2 this requirement can be satisfied through a simple replacement $x(t) = e^{ct} y(t)$.

If the pair of matrices $(A(t), B(t))$ has the property Ω, then the system (3.4.8)–(3.4.10), as a consequence of (1.10.25), simplifies and takes on the form

$$\eta' = [A^B B + (E - A^B A)(YA)'(YA)^{\mathscr{D}}]\eta + (YA)^{\mathscr{D}}(YA)'(E - A^B A)\zeta + A^B f, \qquad (3.5.2)$$

$$(E - A^B A)YA\zeta' = \zeta - u + (E - A^B A)Yf, \tag{3.5.3}$$

$$(E - BY)A\zeta' = (E - BY)f. \tag{3.5.4}$$

A peculiarity of the system (3.5.2)–(3.5.4) is the fact that equation (3.5.3) from it allows for an explicit solution for ζ, and the substitution of this solution into equation (3.5.4) (as a consequence of the equivalence of the systems (3.5.2)–(3.5.4) and (3.4.2)–(3.4.4)) leads to the explicit condition for resolving the initial system

$$A(t)x'(t) = B(t)x(t) + f(t), \quad \alpha \leqslant t \leqslant \beta. \tag{3.5.5}$$

In order to solve equation (3.5.3), we write it as

$$\zeta = (E - A^B A)YA\zeta' + u - (E - A^B A)Yf$$

and construct the iterative process

$$\zeta_{v+1} = T\zeta_v + u - (E - A^B A)Yf,$$

$$\zeta_0 = 0, \quad v = 0, 1, \ldots, \tag{3.5.6}$$

where T is an operator the action of which on the function χ is determined by the formula

$$T\chi = (E - A^B A)YA\chi', \qquad T^0\chi = \chi. \tag{3.5.7}$$

In this case, as a consequence of the constancy of the matrices $(E - A^B A)YA$ and inequalities $A^B AYA = YAA^B A$ and $(A^B A)^2 = A^B A$, for the degrees of the operator (3.5.7) the expression

$$T^v\chi = [(E - A^B A)(YA)^v\chi]^{(v)}, \quad v = 0, 1, \ldots, \tag{3.5.8}$$

is valid, for which, according to property (1.10.25), it follows that if $v \geqslant k$ (k being the matrix index YA), then

$$T^v\chi = 0.$$

This allows us to conclude that, starting from the $(k-1)$th step, the values of the iteration (3.5.6) stabilize and arrive at the solution of equation (3.5.3) which, in view of formula (3.5.8) and the equalities $(E - A^B A)u = (E - A^B A)(E - YB)x = (E - YB)x = u$ (see (1.10.15)), can be written as

$$\zeta = \sum_{i=0}^{k-1} [(YA)^v u - (E - A^B A)(YA)^v Yf]^{(v)} \tag{3.5.9}$$

(it is obvious that since the solution $x \in \mathfrak{A}$, then it is necessary that $\zeta = (E - A^B A)x \in \mathfrak{A}$ and this, when formulating problems for equation (3.5.5), must find its reflection in the requirements on the right-hand side of f).

Next, we note that since the pair of matrices (B, A), according to the condition, is perfect, the equalities

$$(E - BY)A(YA)^v(E - YB) = 0, \quad v = 0, 1, \ldots, \tag{3.5.10}$$

are satisfied for it and, consequently,

$$(E - BY)A(YA)^v u = (E - BY)A(YA)^v(E - YB)x = 0.$$

But then, on substituting the solution (3.5.9) into equation (3.5.4) in view of the constancy of the matrix $(E - BY)A$ and the equality $(E - BY)AA^B = 0$ (see (1.10.16)), we have

$$-\sum_{v=0}^{k-1}[(E - BY)(AY)^{v+1}f]^{(v+1)} = (E - BY)f,$$

which can be transformed into the form

$$\sum_{v=0}^{k}[(E - BY)(AY)^v f]^{(v)} = 0. \tag{3.5.11}$$

Thus, we have obtained the compatibility condition for equation (3.5.5).

Note that since $(E - BY)(AY)^s = 0$, where s is the AY matrix index, the equality (3.5.11) can be given the more perfect form

$$\sum_{v=0}^{r}[(E - BY)(AY)^v f]^{(v)} = 0, \tag{3.5.12}$$

where $r = \min(k, s - 1)$ (see (1.10.8)).

We now make use of the relationship (3.2.6) by applying it to equation (3.5.2). As a result (in view of the equalities $x = A^B Ax + (E - A^B A)x = \eta + \zeta$), for values of any solution of equation (3.5.5) we obtain the relationship:

$$x(t) = X(t)X^{-1}(s)A^B(s)A(s)x(s)$$
$$+ \int_s^t X(t)X^{-1}(\tau)\{[Y(\tau)A(\tau)]^{\mathscr{D}}[Y(\tau)A(\tau)]'\zeta(\tau; u(\tau))$$
$$+ A^B(\tau)f(\tau)\}\,d\tau + \zeta(t; u(t)), \tag{3.5.13}$$

in which $X(t)$ is the solution of the Cauchy problem

$$X'(t) = \{A^B(t)B(t) + [E - A^B(t)A(t)][Y(t)A(t)]'[Y(t)A(t)]^{\mathscr{D}}\}X(t),$$
$$X(\alpha) = E, \tag{3.5.14}$$

where the vector $\zeta(t; u(t))$ is calculated by formula (3.5.9), and $u(t) = [E - Y(t)B(t)]x(t)$.

Using direct substitutions the reader can make sure that something even more important occurs here: it appears that at any vector $u(t) \in \mathfrak{A}$, that satisfies the equation $B(t)u(t) = 0, \alpha \leqslant t \leqslant \beta$, the vector $x(t)$ calculated by formula (3.5.13) is the solution of equation (3.5.5).

Sometimes, however, preference should be given to a different form of representation of the relationship (3.5.13), which can be arrived at by applying the following lemma.

Lemma 3.5.1 The equality

$$X(t)X^{-1}(s)A^B(s)A(s) = A^B(t)A(t)Z(t)Z^{-1}(s), \qquad (3.5.15)$$

is valid, where $X(t)$ and $Z(t)$ are solutions, respectively, of the following Cauchy problems

$$X' = [A^BB + (A^BA)']X, \quad X(\alpha) = E, \qquad (3.5.16)$$

$$Z' = [A^BB - (A^BA)']Z, \quad Z(\alpha) = E. \qquad (3.5.17)$$

Proof By means of a simple substitution on the basis of (3.5.17), (1.10.12), (1.10.21), and (1.10.30), it is easy to see that the matrix

$$T(t, s) = A^B(t)A(t)Z(t)Z^{-1}(s) \qquad (3.5.18)$$

is the solution of the problem

$$T'(t, s) = \{A^B(t)B(t) + [A^B(t)A(t)]'\}T(t, s),$$

$$T(s, s) = A^B(s)A(s).$$

But then, as is known, the matrix $T(t, s)$ is written as

$$T(t, s) = X(t)X^{-1}(s)A^B(s)A(s), \qquad (3.5.19)$$

where $X(t)$ is the solution of the problem (3.5.16). By comparing (3.5.18) with (3.5.19), we obtain (3.5.15). The lemma is thereby proved.

Now, one has to remember that the second summand in braces in (3.5.14) comes from the derivative of the matrix $A^BA^\dagger$ and, taking into account the equalities $(YA)^{\mathscr{D}} = A^BA(YA)^{\mathscr{D}}$ and $A^B = A^BAA^B$, we write a different form of the relationship (3.5.13):

$$x(t) = A^B(t)A(t)Z(t)Z^{-1}(s)x(s) + A^B(t)A(t)\int_s Z(t)Z^{-1}(\tau)$$

$$\times \{[Y(\tau)A(\tau)]^{\mathscr{D}}[Y(\tau)A(\tau)]'\zeta(\tau; u(\tau)) + A^B(\tau)f(\tau)\}\,d\tau + \zeta(t; u(t)), \quad (3.5.20)$$

where $Z(t)$ is the solution of the problem

$$Z'(t) = \{A^B(t)B(t) + [E - A^B(t)A(t)][Y(t)A(t)]'[Y(t)A(t)]^{\mathscr{D}}\}Z(t), \quad (3.5.21)$$

$$Z(\alpha) = E,$$

where the vector $\zeta(t; u(t))$ is calculated by formula (3.5.9), and $u(t) = [E - Y(t)B(t)]x(t)$.

If the matrix A^BA is constant, then the relationships (3.5.13) and (3.5.20) get

---

* It can be supplemented by the derivative $(A^BA)'$, taking into account that in (3.5.2) $\eta = A^BA_\eta = A^BAx$. In this case it appears that the repeated use of Lemma 3.5.1 allows us to put $Z(t) = X(t)$.

simplified considerably. Indeed, in this case

$$(E - A^B A)(YA)'(YA)^{\mathscr{D}} = (YA)'(E - A^B A)(YA)^{\mathscr{D}} = 0,$$
$$(YA)^{\mathscr{D}}(YA)'(E - A^B A) = (YA)^{\mathscr{D}}(E - A^B A)(YA)' = 0.$$

Therefore, for example, the relationship (3.5.20) assumes the form

$$x(t) = A^B(t)A(t)Z(t)Z^{-1}(s)x(s) + A^B(t)A(t) \int_s^t Z(t)Z^{-1}(\tau)A^B(\tau)f(\tau)\,d\tau + \zeta(t; u(t)),$$

where $Z(t)$ is the solution of the problem

$$Z'(t) = A^B(t)B(t)Z(t), \qquad Z(\alpha) = E,$$

and $\zeta(t; u(t))$ and $u(t)$ have the same meaning.

Such a simplification justifies introducing the following definition.

Definition 3.5.2   It will be said that the pair of matrices $(A(t), B(t))$ defined on the segment $\alpha \leqslant t \leqslant \beta$, has the full property Ω on this segment if all the requirements mentioned in Definition 3.5.1 are satisfied for it and, moreover, the matrix $A^B A$ is constant.

The use of these results will be given in the following sections.

3.6.   Solution of the redefined systems

Let us formulate the following problem.

Problem 3.6.1   Let the pair of matrices $(A(t), B(t))$ in the system

$$A(t)x'(t) = B(t)x(t) + f(t), \qquad \alpha \leqslant t \leqslant \beta, \tag{3.6.1}$$

have the property Ω, and the columns of the matrix $B(t)$ be linearly independent. Let, moreover, the matrix Y from the resolving pair of matrices and vector f from the system (3.6.1) be continuously differentiable together with all derivatives up to the $(k-1)$th order, where k is the YA matrix index. It is necessary to find a continuously differentiable solution of the system (3.6.1) that satisfies the condition

$$\int_\alpha^\beta [d\sigma(s)]C(s)x(s) = a, \tag{3.6.2}$$

where a is a given vector.

Before starting to solve this problem, we must note that as a consequence of the constancy of the matrix $(E - A^B A)YA$, the matrix $(E - A^B A)(YA)^\nu$ is also constant (at any $\nu = 1, 2, \ldots$). This follows from the fact that the permutability of the matrices $A^B A$ and YA and the equality $(A^B A)^2 = A^B A$ allows us to write $(E - A^B A)(YS)^\nu = [(E - A^B A)YA]^\nu$.

Moreover, we note that since, according to the condition, the matrix B columns are linearly independent, the equality $u = (E - YB)x = 0$ holds. But then formula (3.5.9), for calculating the vector ζ, can be written as

$$\zeta = -(E - A^B A) \sum_{v=0}^{k-1} (YA)^v (Yf)^{(v)}. \tag{3.6.3}$$

We now substitute the vector (3.5.20) into the condition (3.6.2) (by preliminarily replacing s and t in the relationship (3.5.20). As a result we obtain:

$$\Gamma Z^{-1}(t)x(t) = a + \varphi(t) + \psi, \tag{3.6.4}$$

where

$$\Gamma = \int_\alpha^\beta [d\sigma(s)]C(s)A^B(s)A(s)Z(s), \tag{3.6.5}$$

$$\varphi(t) = \int_\alpha^\beta [d\sigma(s)]C(s)A^B(s)A(s)Z(s)\Phi(s,t), \tag{3.6.6}$$

$$\Phi(s,t) = \int_s^t Z^{-1}(\tau)\{[Y(\tau)A(\tau)]^{\mathscr{D}}[Y(\tau)A(\tau)]'\zeta(\tau) + A^B(\tau)f(\tau)\}\,d\tau, \tag{3.6.7}$$

$$\psi = -\int_\alpha^\beta [d\sigma(s)]C(s)\zeta(s), \tag{3.6.8}$$

the matrix $Z(t)$ is the solution of the Cauchy problem

$$Z'(t) = [A^B B + (E - A^B A)(YA)'(YA)^{\mathscr{D}}]Z, \tag{3.6.9}$$

$$Z(\alpha) = E,$$

and the vector $\zeta(t)$ is calculated by formula (3.6.3).

We now put $s = t$ into the relationship (3.5.20). Then the equality

$$[E - A^B(t)A(t)]x(t) = \zeta(t). \tag{3.6.10}$$

will join the equality (3.6.4).

Thus, the following lemma is valid.

Lemma 3.6.1 Any solution of Problem 3.6.1 satisfies the algebraic system (3.6.10), (3.6.4).

We now fix an arbitrary point $t_0 \in [\alpha, \beta]$ and find the solution of the system (3.6.10), (3.6.4) for $t = t_0$. For this purpose, we first of all show that one of the semi-inverse matrices to the matrix of the system (3.6.10), (3.6.4) is the matrix $T = (E - A^B A^-, A^B AZ\Gamma^-)$.

Indeed, on the basis of Lemma 3.5.1 and the validity of the equality $(A^B A)^2 = A^B A$, one can write

$$\Gamma Z^{-1}(t) = \Gamma_1 X^{-1}(t)A^B(t)A(t) = \Gamma_1 X^{-1}(t)A^B(t)A(t)A^B(t)A(t)$$
$$= \Gamma Z^{-1}(t)A^B(t)A(t), \tag{3.6.11}$$

where

$$\Gamma_1 = \int_\alpha^\beta [d\sigma(s)] C(s) X(s).$$

But then

$$\left(\frac{E - A^B A}{\Gamma Z^{-1}}\right) T \left(\frac{E - A^B A}{\Gamma Z^{-1}}\right) = \left(\frac{E - A^B A}{\Gamma Z^{-1} A^B A Z \Gamma^- \Gamma Z^{-1}}\right) = \left(\frac{E - A^B A}{\Gamma Z^{-1}}\right),$$

and this was to be proved.

Now, taking into account Corollary 1.3.1 and the equality (3.6.11), it is easy to obtain the compatibility condition for the system (3.6.10), (3.6.4)

$$(E - \Gamma\Gamma^-)[a + \varphi(t_0) + \psi] = 0 \tag{3.6.12}$$

(for $t = t_0$) and its general solution

$$x(t_0) = A^B(t_0) A(t_0) Z(t_0) \Gamma^- [a + \varphi(t_0) + \psi] + \zeta(t_0) + \gamma,$$

where γ is an arbitrary solution of the system

$$[E - A^B(t_0) A(t_0)] \gamma = 0, \qquad \Gamma Z^{-1}(t_0) \gamma = 0. \tag{3.6.13}$$

Further reasoning related to the construction of the formula for a general solution of Problem 3.6.1 can be omitted: it would simply be a repetition of the relevant reasoning with respect to the problem (3.2.7), (3.2.8) and Lemma 3.2.1.

Thus, let us formulate the final results.

Theorem 3.6.1   Problem 3.6.1 is solvable if and only if at least at one point $t_0 \in [\alpha, \beta]$ the equality (3.6.12) and, on the entire segment $[\alpha, \beta]$, the identity

$$\sum_{v=0}^r \{[E - B(t) Y(t)][A(t) Y(t)]^v f(t)\}^{(v)} = 0, \qquad \alpha \leqslant t \leqslant \beta,$$

holds true, where $r = \min(k, s - 1)$ $(k \geqslant \mathrm{ind}(YA), s \geqslant \mathrm{ind}(AY))$.

Theorem 3.6.2   The general solution of Problem 3.6.1 is represented as

$$x(t) = A^B(t) A(t) Z(t) Z^{-1}(t_0) x(t_0) + A^B(t) A(t) \int_{t_0}^t Z(t) Z^{-1}(\tau)$$

$$\times \{[Y(\tau) A(\tau)]^{\mathscr{D}} [Y(\tau) A(\tau)]' \zeta(\tau) + A^B(\tau) f(\tau)\} \, d\tau + \zeta(t), \tag{3.6.14}$$

$$x(t_0) = A^B(t_0) A(t_0) Z(t_0) \Gamma^- [a + \varphi(t_0) + \psi] + \zeta(t_0) + \gamma, \tag{3.6.15}$$

where t_0 is an arbitrary fixed point of the segment $[\alpha, \beta]$, $Z(t)$ is the solution of the problem

$$Z'(t) = \{A^B(t) B(t) + [E - A^B(t) A(t)][Y(t) A(t)]' [Y(t) A(t)]^{\mathscr{D}}\} Z(t), \quad Z(\alpha) = E, \tag{3.6.16}$$

the matrix Γ is the matrix (3.6.5), γ is an arbitrary solution of the system (3.6.13), and the vectors $\zeta(t)$, $\varphi(t)$, and ψ are calculated, respectively, by the formulae

$$\zeta(t) = -[E - A^B(t)A(t)]Y(t)f(t) - (E - A^BA)\sum_{v=1}^{k-1}(YA)^v[Y(t)f(t)]^{(v)} \quad (3.6.17)$$

and (3.6.6)–(3.6.8) (remember that the matrices $(E - A^BA)(YA)^v$ ($v = 1, 2, \ldots$) are constant).

Theorem 3.6.3   Problem 3.6.1 has one solution if and only if $\gamma = 0$ is the unique solution of the system (3.6.13).

   If the pair of matrices $(A(t), B(t))$ in Problem 3.6.1 has the full property Ω (see Definition 3.5.2), then the formulation of Theorem 3.6.2 is simplified considerably.

Theorem 3.6.4   A general solution of Problem 3.6.1, in which the pair of matrices (A, B) has the full property Ω, is represented as

$$x(t) = A^BAZ(t)Z^{-1}(t_0)x(t_0) + A^BA\int_{t_0}^t Z(t)Z^{-1}(\tau)A^B(\tau)f(\tau)\,d\tau + \zeta(t),$$

$$x(t_0) = A^BAZ(t_0)\Gamma^-[a + \varphi(t_0) + \psi] + \zeta(t_0) + \gamma, \qquad (3.6.18)$$

where t_0 is an arbitrary fixed point of the segment $[\alpha, \beta]$, $Z(t)$ is the solution of the problem

$$Z'(t) = A^B(t)B(t)Z(t), \qquad Z(\alpha) = E,$$

γ is an arbitrary solution of the system

$$(E - A^BA)\gamma = 0, \qquad \Gamma Z^{-1}(t_0)\gamma = 0,$$

and the matrix Γ and the vectors $\varphi(t)$, ψ, and $\zeta(t)$ are defined by the formulae

$$\Gamma = \int_\alpha^\beta [d\sigma(s)]C(s)A^BAZ(s),$$

$$\varphi(t) = \int_\alpha^\beta [d\sigma(s)]C(s)A^BAZ(s)\int_s^t Z^{-1}(\tau)A^B(\tau)f(\tau)\,d\tau, \qquad (3.6.19)$$

$$\psi = -\int_\alpha^\beta [d\sigma(s)]C(s)\zeta(s),$$

$$\zeta(t) = -(E - A^BA)\sum_{v=0}^{k-1}(YA)^v[Y(t)f(t)]^{(v)} \quad (k \geqslant \mathrm{ind}\,(YA)). \qquad (3.6.20)$$

3.7.   Systems with a regular pair of matrices

Definition 3.7.1   The pair of square matrices $(A(t), B(t))$, defined on the segment $\alpha \leqslant t \leqslant \beta$, is said to be regular on this segment if there exists a constant c at which the matrix $B(t) - cA(t)$ is reversible at all $t \in [\alpha, \beta]$.

It is obvious that if in the system

$$A(t)x'(t) = B(t)x(t) + f(t), \quad \alpha \leqslant t \leqslant \beta,$$

the pair of matrices $(A(t), B(t))$ is regular, then by means of the substitution $x(t) = e^{ct}y(t)$ the system can be brought to a system of the form

$$A(t)x'(t) = x(t) + f(t), \quad \alpha \leqslant t \leqslant \beta. \tag{3.7.1}$$

For that reason, we can confine our attention to considering only the systems having the form (3.7.1).

Definition 3.7.2   It will be said that the matrix $A(t)$, defined on the segment $[\alpha, \beta]$, has on this segment the property Ω if the matrices $A(t)$ and $A^{\mathscr{D}}(t)$ are continuously differentiable and the matrix $(E - A^{\mathscr{D}}A)A$ is constant.

Definition 3.7.3   It will be said that the matrix $A(t)$, defined on the segment $[\alpha, \beta]$, has on this segment the full property Ω if, for it, all the requirements mentioned in Definition 3.7.2 are satisfied and, moreover, the matrix $A^{\mathscr{D}}A$ is constant.

Let us formulate the following problem.

Problem 3.7.1   Let the matrix $A(t)$ in the system (3.7.1) have the property Ω. Let, moreover, the vector f from the system (3.7.1) be continuously differentiable together with its derivatives up to the $(k-1)$th order, where k is the index of the matrix A. Find a continuously differentiable solution of the system (3.7.1) satisfying the condition

$$\int^{\beta} [d\sigma(s)]C(s)x(s) = a, \tag{3.7.2}$$

where a is a given vector.

It is obvious that the conditions of Problem 3.7.1 coincide with those of Problem 3.6.1 when $B(t) \equiv E$; therefore, one can formulate the final results.

Theorem 3.7.1   The general solution of Problem 3.7.1 is represented as

$$x(t) = A^{\mathscr{D}}(t)A(t)Z(t)Z^{-1}(t_0)x(t_0)$$

$$+ A^{\mathscr{D}}(t)A(t) \int_{t_0}^{t} Z(t)Z^{-1}(\tau)[A^{\mathscr{D}}(\tau)A'(\tau)\zeta(\tau) + A^{\mathscr{D}}(\tau)f(\tau)] \, d\tau + \zeta(t), \tag{3.7.3}$$

$$x(t_0) = A^{\mathscr{D}}(t_0)A(t_0)Z(t_0)\Gamma^-[a + \varphi(t_0) + \psi] + \zeta(t_0) + \gamma,$$

where t_0 is an arbitrary fixed point of the segment $[\alpha, \beta]$, $Z(t)$ is the solution of the problem

$$Z'(t) = \{A^{\mathscr{D}}(t) + [E - A^{\mathscr{D}}(t)A(t)]A'(t)A^{\mathscr{D}}(t)\}Z(t), \tag{3.7.4}$$

$$Z(\alpha) = E,$$

γ is an arbitrary solution of the system

$$[E - A^{\mathscr{D}}(t_0)A(t_0)]\gamma = 0, \qquad \Gamma Z^{-1}(t_0)\gamma = 0, \tag{3.7.5}$$

and the matrix Γ and the vectors $\varphi(t)$, ψ, and $\zeta(t)$ are obtained by the formulae

$$\Gamma = \int_\alpha^\beta [d\sigma(s)] A^{\mathscr{D}}(s) Z(s), \tag{3.7.6}$$

$$\varphi(t) = \int_\alpha^\beta [d\sigma(s)] C(s) A^{\mathscr{D}}(s) A(s) Z(s) \Phi(s, t), \tag{3.7.7}$$

$$\Phi(s, t) = \int_s^t Z^{-1}(\tau) [A^{\mathscr{D}}(\tau) A'(\tau)\zeta(\tau) + A^{\mathscr{D}}(\tau) f(\tau)] d\tau, \tag{3.7.8}$$

$$\psi = - \int_\alpha^\beta [d\sigma(s)] C(s)\zeta(s), \tag{3.7.9}$$

$$\zeta(t) = - [E - A^{\mathscr{D}}(t)A(t)] f(t) - (E - A^{\mathscr{D}}A) \sum_{v=1}^{k-1} A^v f^{(v)}(t), \tag{3.7.10}$$

where $k \geqslant \operatorname{ind} A(t)$ and the matrices $(E - A^{\mathscr{D}}A)A^v$ $(v = 1, 2, \ldots)$ are constant.

Theorem 3.7.2 Problem 3.7.1 is solvable if and only if at least at one point $t_0 \in [\alpha, \beta]$ the equality

$$(E - \Gamma\Gamma^-)[a + \varphi(t_0) + \psi] = 0 \tag{3.7.11}$$

is satisfied.

Theorem 3.7.3 Problem 3.7.1 has only one solution if and only if $\gamma = 0$ is the unique solution of the system (3.7.5).

Theorem 3.7.4 The general solution of Problem 3.7.1, in which the matrix has the full property Ω, is represented as

$$x(t) = A^{\mathscr{D}} A Z(t) Z^{-1}(t_0) x(t_0) + A^{\mathscr{D}} A \int_{t_0}^t Z(t) Z^{-1}(\tau) A^{\mathscr{D}}(\tau) f(\tau) d\tau + \zeta(t),$$

$$x(t_0) = A^{\mathscr{D}} A Z(t_0) \Gamma^- [a + \varphi(t_0) + \psi] + \zeta(t_0) + \gamma, \tag{3.7.12}$$

where t_0 is an arbitrary fixed point of the segment $[\alpha, \beta]$, $Z(t)$ is the solution of the problem

$$Z'(t) = A^{\mathscr{D}}(t) Z(t), \qquad Z(\alpha) = E, \tag{3.7.13}$$

γ is an arbitrary solution of the problem

$$(E - A^{\mathscr{D}}A)\gamma = 0, \qquad \Gamma Z^{-1}(t_0)\gamma = 0, \tag{3.7.14}$$

and the matrix Γ and the vectors $\varphi(t), \psi$, and $\zeta(t)$ are defined by the formulae

$$\Gamma = \int_\alpha^\beta [d\sigma(s)]C(s)A^\mathscr{D}AZ(s), \tag{3.7.15}$$

$$\varphi(t) = \int_\alpha^\beta [d\sigma(s)]C(s)A^\mathscr{D}A \int_s^t Z(s)Z^{-1}(\tau)A^\mathscr{D}(\tau)f(\tau)d\tau, \tag{3.7.16}$$

$$\psi = -\int_\alpha^\beta [d\sigma(s)]C(s)\zeta(s), \tag{3.7.17}$$

$$\zeta(t) = -(E - A^\mathscr{D}A) \sum_{v=0}^{k-1} A^v f^{(v)}(t), \tag{3.7.18}$$

where $k \geqslant \operatorname{ind} A(t)$ and the matrices $(E - A^\mathscr{D}A)A^v$ $(v = 0, 1, 2, \ldots)$ are constant.

Remark 3.7.1   Note that the vector $\zeta = (E - A^\mathscr{D}A)x$ (see (3.7.10), (3.7.18)) is defined uniquely from the right-hand side of f (of the system being solved); therefore additional conditions, which it must satisfy, are not needed.

In this connection, it must be considered that the condition (3.7.2) in Problem 3.7.1 is natural when the vector ζ in φ and ψ (see (3.7.7)–(3.7.9) and (3.7.16), (3.7.17)) is not involved. In particular, such will be the case when the matrix $C(s)$, from condition (3.7.2), satisfies the system

$$C(s)[E - A^\mathscr{D}(s)A(s)] = 0, \tag{3.7.19}$$

$$C(s)Z(s)Z^{-1}(\tau)A^\mathscr{D}(\tau)A'(\tau)[E - A^\mathscr{D}(\tau)A(\tau)] = 0, \quad \alpha \leqslant s \leqslant \beta, \alpha \leqslant \tau \leqslant \beta \tag{3.7.20}$$

Obviously the solution of the system (3.7.19), (3.7.20) has the form

$$C(s) = \mathscr{D}(s)A^\mathscr{D}(s)A(s). \tag{3.7.21}$$

On substituting (3.7.21) into equation (3.7.20), and in this case taking into account Lemma 3.5.1, for the matrix $\mathscr{D}(s)$ we obtain:

$$\mathscr{D}(s)X(s)X^{-1}(\tau)A^\mathscr{D}(\tau)A'(\tau)[E - A^\mathscr{D}(\tau)A(\tau)] = 0, \tag{3.7.22}$$

where $X(t)$ is the solution of the problem

$$X'(t) = \{A^\mathscr{D}(t) + [E - A^\mathscr{D}(t)A(t)]A'(t)A^\mathscr{D}(t)\}X(t),$$

$$X(\alpha) = E.$$

By solving equation (3.7.22) with the help of Lemma 1.3.3, we obtain:

$$\mathscr{D}(s) = G(s)(E - RR^-)X^{-1}(s), \tag{3.7.23}$$

where

$$R = \int_\alpha^\beta T(\tau)T^*(\tau)d\tau, \qquad T(\tau) = X^{-1}(\tau)A^\mathscr{D}(\tau)A'(\tau)[E - A^\mathscr{D}(\tau)A(\tau)],$$

and $G(s)$ is an arbitrary matrix.

On substituting now (3.7.23) into (3.7.21) we obtain:

$$C(s) = G(s)(E - RR^-)X^{-1}(s)A^{\mathscr{D}}(s)A(s). \tag{3.7.24}$$

Formula (3.7.24) gives a set of matrices $C(s)$, for which condition (3.7.2) is natural.

If the matrix $A(t)$ has the full property Ω, then formula (3.7.24) simplifies and assumes the form

$$C(s) = G(s) - A^{\mathscr{D}}(s)A(s) \tag{3.7.25}$$

where $G(s)$ is an arbitrary matrix. This is because in such a case $R = 0$ and, moreover, because from the arbitrary matrices, $G(s)$ the arbitrariness of the matrix $G(s)X^{-1}(s)$ follows.

Note that unnatural conditions (in the sense defined here) can turn out to be natural because of other reasons: for example, quite justified conditions arise in the case of minimization of a certain functional in connection with the construction of approximate methods of solving equations with partial derivatives. As an example, we consider the system

$$Ax' = x + f, \quad 0 \leqslant t \leqslant H, \tag{3.7.26}$$

$$A = - \begin{pmatrix} 0 & \Sigma_a^{-1} & 0 \\ (3\Sigma)^{-1} & 0 & 2(3\Sigma)^{-1} \\ 0 & 2(5\Sigma)^{-1} & 0 \end{pmatrix},$$

in which Σ_a and Σ are constant and $0 < \Sigma_a \leqslant \Sigma$. The system (3.7.26) is a P_2-approximation of the transfer equation of neutrons in a plane-parallel stationary case (for details see [27, p. 127]).

Conditions for the system (3.7.26) are constructed on the basis of the requirement of a rather good approximation of the accurate conditions

$$\sum_{i=0}^{\infty} (2i + 1)x_i(0)P_i(\mu) = 0, \quad 0 \leqslant \mu \leqslant 1,$$

$$\sum_{i=0}^{\infty} (2i + 1)x_i(H)P_i(\mu) = 0, \quad -1 \leqslant \mu < 0, \tag{3.7.27}$$

where $P_i(\mu)$ are Legendre polynomials. In the case of a P_2-approximation, the equalities (3.7.27) are replaced by the equalities

$$(P_0(\mu)3P_1(\mu)5P_2(\mu))x(0) = 0, \quad 0 \leqslant \mu \leqslant 1, \tag{3.7.28}$$

$$(P_0(\mu)3P_1(\mu)5P_2(\mu))x(H) = 0, \quad -1 \leqslant \mu < 0. \tag{3.7.29}$$

It is obvious that it is not possible to satisfy conditions (3.7.28), (3.7.29) and, for example, the Vladimirov–Marshak conditions

$$(4 \quad 8 \quad 1)x(0) = 0, \quad (4 \quad -8 \quad 1)x(H) = 0 \tag{3.7.30}$$

are obtained by multiplying (3.7.28) and (3.7.29) by μ and by subsequently integrating the equality (3.7.28) from 0 to 1 and the equality (3.7.29) from -1 to 0.

Another method of obtaining conditions for the system (3.7.26) implies that the equalities (3.7.28) and (3.7.29) are fixed at some point μ. For example, if we put $\mu = 3/5$ in (3.7.28) and $\mu = -3/5$ in (3.7.29), we then obtain the conditions

$$(1 \quad 3\sqrt{3/5} \quad 2)x(0) = 0, \qquad (1 \quad -3\sqrt{3/5} \quad 2)x(H) = 0. \qquad (3.7.31)$$

In order to obtain the matrices $C(s)$ involved in (3.7.2) corresponding to the conditions (3.7.30) and (3.7.31), we can use the linear interpolation. In the case (3.7.30) and (3.7.31) we then have

$$C(s) = \left(4, 8 - \frac{16}{H}s, 1\right) \qquad (3.7.32)$$

and

$$C(s) = \left(1, 3\sqrt{3/5}\left(1 - \frac{2}{H}s\right), 2\right), \qquad (3.7.33)$$

respectively.

As the matrix $\sigma(s)$ in condition (3.7.2) one should take a matrix corresponding to a two-point boundary-value problem.

In order to answer the question of the natural character of the conditions with the matrices (3.7.32), (3.7.33), it suffices to verify the validity of the equality

$$C(s) = C(s)A^{\mathscr{D}}A, \qquad (3.7.34)$$

because it is obvious that this equality is necessary and sufficient for representation of the matrix $C(s)$ in the form of (3.7.25).

The matrix $A^{\mathscr{D}}$ can be obtained with the help of Theorem 1.9.1. In this case it appears that

$$A^{\mathscr{D}} = \frac{15\Sigma_a\Sigma^2}{5\Sigma + 4\Sigma_a}A.$$

But then

$$A^{\mathscr{D}}A = \frac{1}{4\Sigma_a + 5\Sigma}\begin{pmatrix} 5\Sigma & 0 & 10\Sigma \\ 0 & 4\Sigma_a + 5\Sigma & 0 \\ 2\Sigma_a & 0 & 4\Sigma_a \end{pmatrix}.$$

It is now easy to make sure that for the matrix (3.7.32) $C(s) \neq C(s)A^{\mathscr{D}}A$ and for the matrix (3.7.33) $C(s) = C(s)A^{\mathscr{D}}A$, i.e. conditions (3.7.31) are natural and conditions (3.7.30) are unnatural.

To conclude, we note that all conditions of the form

$$(1 \quad \alpha \quad 2)x(0) = 0, \qquad (1 \quad -\alpha \quad 2)x(H) = 0$$

for the system (3.7.26) are natural, which can be verified through a direct verification of the equality (3.7.34).

3.8. Systems with a matrix of index 2

If ind $A(t) \leqslant 2$, then the system (3.4.8)–(3.4.10), which is equivalent to the equation

$$A(t)x'(t) = x(t) + f(t), \quad \alpha \leqslant t \leqslant \beta, \tag{3.8.1}$$

has the form

$$\eta' = [A^{\mathcal{D}} + (E - A^{\mathcal{D}}A)AA'(A^{\mathcal{D}})^2 + (E - A^{\mathcal{D}}A)A'A^{\mathcal{D}}]\eta$$
$$+ [(A^{\mathcal{D}})^2 A'A(E - A^{\mathcal{D}}A) + A^{\mathcal{D}}A'(E - A^{\mathcal{D}}A)]\zeta + A^{\mathcal{D}}f, \tag{3.8.2}$$
$$(E - A^{\mathcal{D}}A)A\zeta' = \zeta - (E - A^{\mathcal{D}}A)AA'A^{\mathcal{D}}\eta + (E - A^{\mathcal{D}}A)f. \tag{3.8.3}$$

A distinctive feature of the system (3.8.2), (3.8.3) is the fact that equation (3.8.3) can be reduced to an algebraic system with respect to ζ. Indeed, by multiplying equation (3.8.3) by the matrix A, and in view of the inequality ind $A(t) \leqslant 2$, we obtain:

$$A\zeta = -(E - A^{\mathcal{D}}A)Af. \tag{3.8.4}$$

But in such a case

$$A\zeta' = [A\zeta]' - A'\zeta = -[(E - A^{\mathcal{D}}A)Af]' - A'\zeta$$
$$= (A^{\mathcal{D}}A)'Af - (E - A^{\mathcal{D}}A)(Af)' - A'\zeta,$$

and this, using formula (1.8.13), can be written as

$$A\zeta' = [(E - A^{\mathcal{D}}A)AA'A^{\mathcal{D}} + A^{\mathcal{D}}A'A(E - A^{\mathcal{D}}A)$$
$$- (E - A^{\mathcal{D}}A)A'(E - A^{\mathcal{D}}A)]f - (E - A^{\mathcal{D}}A)Af' - A'\zeta. \tag{3.8.5}$$

On substituting (3.8.5) into the left-hand side of (3.8.3), for the vector ζ we obtain the equation

$$[E + (E - A^{\mathcal{D}}A)A']\zeta = (E - A^{\mathcal{D}}A)AA'A^{\mathcal{D}}(\eta + f)$$
$$- [E + (E - A^{\mathcal{D}}A)A'](E - A^{\mathcal{D}}A)f - (E - A^{\mathcal{D}}A)Af'. \tag{3.8.6}$$

Thus, the solution of equation (3.8.3) is brought to the solution of the system (3.8.4), (3.8.6).

The solution of the system (3.8.4), (3.8.6) is convenient to seek in the form

$$\zeta - y - (E - A^{\mathcal{D}}A)f, \tag{3.8.7}$$

because, for y, we obtain a simpler system, namely

$$Ay = 0, \tag{3.8.8}$$

$$[E + (E - A^{\mathscr{D}}A)A']y = (E - A^{\mathscr{D}}A)AA'A^{\mathscr{D}}(\eta + f) - (E - A^{\mathscr{D}}A)Af'. \tag{3.8.9}$$

If the representation (3.8.7) is substituted into (3.8.2), then as a consequence of (3.8.8) we obtain a different, simpler form of equation (3.8.2):

$$\eta' = A^{\mathscr{D}} + (E - A^{\mathscr{D}}A)AA'(A^{\mathscr{D}})^2 + (E - A^{\mathscr{D}}A)A'A^{\mathscr{D}}]\eta$$
$$+ A^{\mathscr{D}}A'y - A^{\mathscr{D}}[A^{\mathscr{D}}A'A + A'](E - A^{\mathscr{D}}A)f + A^{\mathscr{D}}f. \tag{3.8.10}$$

We are interested in the case when, for the vector y, the system (3.8.8), (3.8.9) has a unique solution: in this case, by solving the system (3.8.8), (3.8.9) and substituting the obtained solution into equation (3.8.10), we arrive at an explicit equation for η, which allows us to find an explicit expression for η and, on substituting η into the formula for y as well as into the compatibility conditions (3.8.8), (3.8.9), to obtain for the initial system (3.8.1) the conditions of compatibility and the formula for the solution

$$x = \eta + y - (E - A^{\mathscr{D}}A)f. \tag{3.8.11}$$

In order to shorten the subsequent calculations, we write the system (3.8.8), (3.8.9) and equation (3.8.10) as

$$Dy = \mathscr{E}\eta + \varphi, \tag{3.8.12}$$

$$\eta' = M\eta + A^{\mathscr{D}}A'y + \psi, \tag{3.8.13}$$

where

$$D = \begin{pmatrix} A \\ E + (E - A^{\mathscr{D}}A)A' \end{pmatrix}, \qquad \mathscr{E} = \begin{pmatrix} 0 \\ (E - A^{\mathscr{D}}A)AA'A^{\mathscr{D}} \end{pmatrix},$$

$$\varphi = \begin{pmatrix} 0 \\ (E - A^{\mathscr{D}}A)AA'A^{\mathscr{D}}f - (E - A^{\mathscr{D}}A)Af' \end{pmatrix},$$

$$M = A^{\mathscr{D}} + (E - A^{\mathscr{D}}A)AA'(A^{\mathscr{D}})^2 + (E - A^{\mathscr{D}}A)A'A^{\mathscr{D}}, \tag{3.8.14}$$

$$\psi = - A^{\mathscr{D}}(A^{\mathscr{D}}A'A + A')(E - A^{\mathscr{D}}A)f + A^{\mathscr{D}}f.$$

Under the assumption that equation (3.8.12) has a unique solution, we obtain:

$$y = F\eta + \theta, \tag{3.8.15}$$

where $F = (D^*D)^{-1}D^*\mathscr{E}$ and $\theta = F \cdot \varphi$.

On substituting (3.8.15) into (3.8.12) and (3.8.13) we arrive at the compatibility condition

$$R\eta = \xi \tag{3.8.16}$$

and at the explicit equation for η:

$$\eta' = N\eta + \chi, \tag{3.8.17}$$

where $R = DF - \mathscr{E}, \xi = \varphi - D\theta, N = M + A^{\mathscr{D}}A'F$, and $\chi = A^{\mathscr{D}}A'\theta + \psi$.

Remember that now $\eta = A^{\mathscr{D}}Ax$ and, consequently, $(E - A^{\mathscr{D}}A)\eta = 0$. But then the compatibility condition (3.8.16) must be supplemented with the equality $(E - A^{\mathscr{D}}A)\eta = 0$ and be written as

$$S\eta = n, \tag{3.8.18}$$

where

$$S = \begin{pmatrix} R \\ E - A^{\mathscr{D}}A \end{pmatrix}, \qquad n = \begin{pmatrix} \xi \\ 0 \end{pmatrix}.$$

If, moreover, there is the condition

$$\int_{\alpha}^{\beta} [d\sigma(s)]C(s)x(s) = a, \tag{3.8.19}$$

then it must be taken into consideration that from (3.8.11) and (3.8.15) the equality

$$x = T\eta + \delta \tag{3.8.20}$$

follows, where $T = E + F$ and $\delta = \theta - (E - A^{\mathscr{D}}A)f$. Substitution of the vector (3.8.20) into (3.8.19) yields in this case the condition for η, namely

$$\int_{\alpha}^{\beta} [d\sigma(s)]C(s)T(s)\eta(s) = b,$$

where

$$b = a - \int_{\alpha}^{\beta} [d\sigma(s)]C(s)\delta(s).$$

Thus, the solution of the problem (3.8.1), (3.8.19) is reduced to solving the following problem for η:

$$\eta'(t) = N(t)\eta(t) + \chi(t), \qquad S(t)\eta(t) = n(t),$$
$$\int_{\alpha}^{\beta} [d\sigma(s)]C(s)T(s)\eta(s) = b \tag{3.8.21}$$

(the solution of the initial problem (3.8.1), (3.8.19) is obtained by formula (3.8.20).

Note that, as for problem (3.8.21), one can quite justifiably say that it is already solved: construction of its solution totally coincides with analogous construction in the case of a perfect group of three matrices (see Section 3.3, starting from Lemma 3.3.3). However, the solution obtained in this case is very unwieldy and is hardly amenable to comprehension: all constituents of the solution turn out to be essential (nothing is cancelled and set to zero). We therefore, omit the formulations of the final results and proceed to consider those particular cases when everything is representable in completed form.

3.9.   Systems with a matrix of index 2. Particular cases

Let us formulate the following problem.

Problem 3.9.1   Let the matrix $A(t)$ and the vector $f(t)$ in the system

$$A(t)x'(t) = x(t) + f(t), \quad \alpha \leqslant t \leqslant \beta, \tag{3.9.1}$$

be such that $A(t) \leqslant 2$, $(A^{\mathscr{D}})' \in \mathfrak{N}, A'' \in \mathfrak{N}$, and $f'' \in \mathfrak{N}$. Moreover, on the entire segment $[\alpha, \beta]$ let the equality

$$(E - A^{\mathscr{D}}A)AA'A^{\mathscr{D}} = 0 \tag{3.9.2}$$

be valid, and let the matrix

$$A^*A + [E + (E - A^{\mathscr{D}}A)A']^*[E + (E - A^{\mathscr{D}}A)A'] \tag{3.9.3}$$

be reversible.

Find a continuously differentiable vector $x(t)$ that satisfies the condition

$$\int_\alpha^\beta [d\sigma(s)]C(s)x(s) = a \tag{3.9.4}$$

and the system (3.9.1).

In order to solve this problem, we use the system (3.8.2), (3.8.3) which in this case has the form

$$\eta' = [A^{\mathscr{D}} + (E - A^{\mathscr{D}}A)A'A^{\mathscr{D}}]\eta + [(A^{\mathscr{D}})^2 A'A(E - A^{\mathscr{D}}A)$$
$$+ A^{\mathscr{D}}A'(E - A^{\mathscr{D}}A)]\zeta + A^{\mathscr{D}}f,$$
$$(E - A^{\mathscr{D}}A)A\zeta' = \zeta + (E - A^{\mathscr{D}}A)f.$$

In exactly the same way as in the preceding section, by representing the solution ζ as the difference

$$\zeta = y - (E - A^{\mathscr{D}}A)f \tag{3.9.5}$$

for the vector y we obtain:

$$Ay = 0, \tag{3.9.6}$$

$$[E + (E - A^{\mathscr{D}}A)A']y = -(E - A^{\mathscr{D}}A)Af'. \tag{3.9.7}$$

If the assumption about the nonsingularity of the matrix (3.9.3) is taken into account, then the solution of the system (3.9.6), (3.9.7) (if it exists) can be represented as

$$y = -D[E + (E - A^{\mathscr{D}}A)A']^* (E - A^{\mathscr{D}}A)Af', \tag{3.9.8}$$

where

$$D = \{A^*A + [E + (E - A^{\mathscr{D}}A)A']^* [E + (E - A^{\mathscr{D}}A)A']\}^{-1}. \tag{3.9.9}$$

On substituting (3.9.8) into (3.9.6) and (3.9.7), we obtain the conditions of compatibility of the system (3.9.6), (3.9.7):

$$AD[E + (E - A^{\mathscr{D}}A)A']^* (E - A^{\mathscr{D}}A)Af' = 0, \tag{3.9.10}$$

$$[E + (E - A^{\mathscr{D}}A)A']D[E + (E - A^{\mathscr{D}}A)A']^* (E - A^{\mathscr{D}}A)Af' = (E - A^{\mathscr{D}}A)Af'. \tag{3.9.11}$$

Next, using the methods applied in Sections 3.2 and 3.7, it is easy to obtain the following results.

Lemma 3.9.1 Any solution of Problem 3.9.1 satisfies the algebraic system

$$[E - A^{\mathscr{D}}(t)A(t)]x(t) = \zeta(t), \qquad \Gamma Z^{-1}(t)x(t) = a + \varphi(t) + \psi,$$

where $Z(t)$ is the solution of the problem

$$Z' = [A^{\mathscr{D}} + (E - A^{\mathscr{D}}A)A'A^{\mathscr{D}}]Z, \qquad Z(\alpha) = E, \tag{3.9.12}$$

$$\Gamma = \int_{\alpha}^{\beta} [d\sigma(s)]C(s)A^{\mathscr{D}}(s)A(s)Z(s), \tag{3.9.13}$$

$$\varphi(t) = \int_{\alpha}^{\beta} [d\sigma(s)]C(s)A^{\mathscr{D}}(s)A(s)Z(s)\Phi(s, t), \tag{3.9.14}$$

$$\Phi(s, t) = \int_{s}^{t} Z^{-1}(\tau)\{A^{\mathscr{D}}(\tau)[A^{\mathscr{D}}(\tau)A'(\tau)A(\tau) + A'(\tau)]\zeta(\tau) + A^{\mathscr{D}}(\tau)f(\tau)\}d\tau, \tag{3.9.15}$$

$$\psi = -\int_{\alpha}^{\beta} [d\sigma(s)]C(s)\zeta(s),$$

and the vector $\zeta(t)$ is obtained by formulae (3.9.5) and (3.9.8).

We note immediately that the condition (3.9.4) is then natural when the vector $\zeta(t)$ in formulae (3.9.14) and (3.9.15) is cancelled. In this case, formulae (3.7.21) and (3.7.23), giving the expression for the matrices $C(s)$, for which the condition (3.9.4) is natural, in this case have the form

$$C(s) = U(s)A^{\mathscr{D}}(s)A(s), \qquad U(s) = G(s)(E - RR^-)X^{-1}(s),$$

where

$$R = \int_{\alpha}^{\beta} T(\tau)T^*(\tau)d\tau,$$

$$T(\tau) = X^{-1}(\tau)A^{\mathscr{D}}(\tau)[A^{\mathscr{D}}(\tau)A'(\tau)A(\tau) + A'(\tau)][E - A^{\mathscr{D}}(\tau)A(\tau)],$$

$X(\tau)$ is the solution of the problem

$$X' = [A^{\mathscr{D}} + (E - A^{\mathscr{D}}A)A'A^{\mathscr{D}}]X, \qquad X(\alpha) = E,$$

and $G(s)$ is an arbitrary matrix. In particular, if the matrix $A^{\mathscr{D}}A$ is constant, then the arbitrary matrix in (3.9.16) is the matrix $U(s)$ (cf. (3.7.25)).

Theorem 3.9.1 The general solution of Problem 3.9.1 is represented as

$$x(t) = A^{\mathscr{D}}(t)A(t)Z(t)Z^{-1}(t_0)x(t_0) + A^{\mathscr{D}}(t)A(t)\int_{t_0}^{t} Z(t)Z^{-1}(\tau)A^{\mathscr{D}}(\tau)$$

$$\times \{[A^{\mathscr{D}}(\tau)A'(\tau)A(\tau) + A'(\tau)]\zeta(\tau) + f(\tau)\}d\tau + \zeta(t),$$

$$x(t_0) = A^{\mathscr{D}}(t_0)A(t_0)Z(t_0)\Gamma^{-}[a + \varphi(t_0) + \psi] + \zeta(t_0) + \gamma, \qquad (3.9.16)$$

where $Z(t)$ is the solution of Problem (3.9.12), t_0 is an arbitrary fixed point of the segment $[\alpha, \beta]$, γ is an arbitrary solution of the system

$$[E - A^{\mathscr{D}}(t_0)A(t_0)]\gamma = 0, \qquad \Gamma Z^{-1}(t_0)\gamma = 0, \qquad (3.9.17)$$

and the matrix Γ and the vectors $\varphi(t)$, ψ, and $\zeta(t)$ are obtained by formulae (3.9.13)–(3.9.15) and (3.9.5), (3.9.8).

Theorem 3.9.2 Problem 3.9.1 has only one solution if and only if the identities (3.9.10) and (3.9.11) hold true and at least at one point $t_0 \in [\alpha, \beta]$ the equality

$$(E - \Gamma\Gamma^{-})[a + \varphi(t_0) + \psi] = a$$

is valid.

Theorem 3.9.3 Problem 3.9.1 has only one solution if and only if $\gamma = 0$ is the unique solution of the system (3.9.17).

The condition (3.9.2) of Problem 3.9.1 is satisfied, in particular, when either the matrix $(E - A^{\mathscr{D}}A)A$ or the matrix $A^{\mathscr{D}}A$ is constant. The first case was considered in Section 3.7 and is now of no special interest. As far as the second case is concerned, we should examine it in greater detail. The point here is that if the matrix $A^{\mathscr{D}}A$ is constant, then as a consequence of the equalities $A^{\mathscr{D}}AA' = A'A^{\mathscr{D}}A$ and $Ay = 0$ the system (3.9.6), (3.9.7) simplifies to assume the form

$$Ay = 0,$$

$$(E + A')y = -(E - A^{\mathscr{D}}A)Af'.$$

At the same time, the matrix (3.9.3), the nonsingularity of which is assumed in Problem 3.9.1, also becomes simpler:

$$A^*A + (E + A')^*(E + A'). \qquad (3.9.18)$$

Simplifications also occur in formulae (3.9.8)–(3.9.14), (3.9.16), and (3.9.17). Therefore, it is appropriate to formulate such a problem.

Problem 3.9.2 Let the matrix $A(t)$ and the vector $f(t)$ in the system (3.9.1) be such that ind $A(t) \leqslant 2$, $(A^{\mathscr{D}})' \in \mathfrak{N}$, $A'' \in \mathfrak{N}$, and $f'' \in \mathfrak{N}$. Moreover, let the matrix $A^{\mathscr{D}}A$ be constant, and let the matrix (3.9.18) be reversible (on the entire segment $[\alpha, \beta]$). It is necessary to find the continuously differentiable vector $x(t)$ satisfying condition (3.9.4) and the system (3.9.1).

The following theorems, which are corollaries of Theorems 3.9.1–3.9.3, hold.

Theorem 3.9.4 The general solution of Problem 3.9.2 is representable as

$$x(t) = A^{\mathscr{D}}AZ(t)Z^{-1}(t_0)x(t_0) + A^{\mathscr{D}}A \int_{t_0}^{t} Z(t)Z^{-1}(\tau)A^{\mathscr{D}}(\tau)f(\tau)d\tau + \zeta(t),$$

$$x(t_0) = A^{\mathscr{D}}AZ(t_0)\Gamma^-[a + \varphi(t_0) + \psi] + \zeta(t_0) + \gamma,$$

where t_0 is an arbitrary fixed point of the segment $[\alpha, \beta]$, $Z(t)$ is the solution of the problem

$$Z'(t) = A^{\mathscr{D}}(t)Z(t), \qquad Z(\alpha) = E,$$

γ is an arbitrary solution of the system

$$(E - A^{\mathscr{D}}A)\gamma = 0, \qquad \Gamma Z^{-1}(t_0)\gamma = 0, \tag{3.9.19}$$

and the matrix Γ and the vectors $\varphi(t)$, ψ, and $\zeta(t)$ are obtained from the formulae

$$\Gamma = \int_{\alpha}^{\beta} [d\sigma(s)]C(s)A^{\mathscr{D}}AZ(s),$$

$$\varphi(t) = \int_{\alpha}^{\beta} [d\sigma(s)]C(s)A^{\mathscr{D}}AZ(s)\Phi(s, t),$$

$$\Phi(s, t) = \int_{s}^{t} Z^{-1}(\tau)A^{\mathscr{D}}(\tau)f(\tau)d\tau, \quad \psi = -\int_{\alpha}^{\beta} [d\sigma(s)]C(s)\zeta(s),$$

$$\zeta(t) = -D(t)[E + A'(t)]^*(E - A^{\mathscr{D}}A)A(t)f'(t) - (E - A^{\mathscr{D}}A)f(t),$$

where

$$D(t) = \{A^*(t)A(t) + [E + A'(t)]^*[E + A'(t)]\}^{-1}.$$

Theorem 3.9.5 Problem 3.9.2 is solvable if and only if the identities

$$A(t)D(t)[E + A'(t)]^*(E - A^{\mathscr{D}}A)A(t)f'(t) = 0,$$

$$[E + A'(t)]D(t)[E + A'(t)]^*(E - A^{\mathscr{D}}A)A(t)f'(t) = (E - A^{\mathscr{D}}A)A(t)f'(t)$$

are valid and at least at one point $t_0 \in [\alpha, \beta]$ the equality

$$(E - \Gamma\Gamma^-)[a + \varphi(t_0) + \psi] = 0$$

holds.

Theorem 3.9.6   Problem 3.9.2 has only one solution if and only if $\gamma = 0$ is the unique solution of the system (3.9.19).

Remark 3.9.1   If the matrix $A^{\mathscr{D}}A$ is constant and $C(s)A^{\mathscr{D}}A = C(s)$, then, as has already been pointed out, the condition (3.9.4) in Problem 3.9.1 turns out to be natural. In this case $\psi = 0$ and the formulae for calculating Γ and $\varphi(t)$ simplify slightly.

3.10.   Stability of linear combinations of the components of solutions of singular systems

Investigating the stability of certain linear combinations of components of the solutions of the singular system

$$A(t)x'(t) = B(t)x(t) + f(t), \quad \alpha \leqslant t \leqslant \beta, \tag{3.10.1}$$

is at least required because the condition

$$\int_\alpha^\beta [d\sigma(s)]C(s)x(s) = a, \tag{3.10.2}$$

which must be satisfied by the solutions of the problems examined in the preceding sections, is actually the requirement to the linear combinations $y(t) = C(t)x(t)$. The accuracy of the numerical solution of the problem of the form (3.10.1), (3.10.2) largely depends on the degree of stability of these combinations.

One of the possibilities of investigating the stability of linear combinations is provided by the following lemma.

Lemma 3.10.1   If the matrices $A(t)$, $B(t)$, and $M(t)$ are such that $A^*M^* + MA \leqslant 0, (MB)^* = MB$, and the matrix MB is constant, then for all solutions of the equation

$$Ax' = Bx, \tag{3.10.3}$$

when $t \geqslant s$, the estimation

$$(MBx(t), x(t)) \leqslant (MBx(s), \quad x(s)) \tag{3.10.4}$$

holds.

Proof   Multiply equation (3.10.3) by the matrix M and then multiply scalarly the obtained product by the vector $2x'$. As a result, we arrive at the equality

$$2(MAx', x') = 2(MBx, x'),$$

from which it obviously follows that

$$2\text{Re}(MAx', x') = 2\,\text{Re}(MBx, x'). \tag{3.10.5}$$

Taking into account the self-conjugacy of the matrix MB, the right-hand side of the equality (3.10.5) can be transformed, namely

$$2\,\text{Re}(MBx, x') = (MBx, x') + (x', MBx) = (MBx, x') + (MBx', x) = (MBx, x)'.$$

Moreover,

$$2\,\text{Re}(MAx', x') = (MAx', x') + (x', MAx')$$

$$= (MAx', x) + (A^*M^*x', x') = ((A^*M^* + MA)x', x').$$

But then the equality (3.10.5) assumes the form

$$((A^*M^* + MA)x', x') = (MBx, x)'. \tag{3.10.6}$$

Note now that if the left-hand side of equality (3.10.6) is nonpositive, which occurs, for example, when $A^*M^* + MA \leqslant 0$, then

$$(MBx, x)' \leqslant 0. \tag{3.10.7}$$

By integrating the inequality (3.10.7) in the limits from s to t, we obtain:

$$(MBx(t), x(t)) \leqslant (MBx(s), x(s)).$$

Lemma 3.10.1 is thereby proved.

We now limit ourselves to considering systems of the form

$$A(t)x'(t) = x(t) + f(t), \quad \alpha \leqslant t \leqslant \beta, \tag{3.10.8}$$

and formulate the following useful statement which is the corollary of Lemma 3.10.1.

Corollary 3.10.1 If the matrix C^*C is constant and $A^*C^*C + C^*CA \leqslant 0$, then for all solutions of the equation

$$A(t)x'(t) = x(t) \tag{3.10.9}$$

for $t \geqslant s$, the estimation

$$\|C(t)x(t)\| \leqslant \|C(s)x(s)\| \tag{3.10.10}$$

holds.

Proof Putting $M = C^*C$ and $B = E$ into the formulation of Lemma 3.10.1, we arrive at the inequality (3.10.10).

Suppose now that the matrix $A(t)$ in the system (3.10.9) satisfies the conditions of at least one of Problems 3.6.1, 3.7.1, 3.9.1, and 3.9.2 or the conditions of the general problem of index 2 (see Section 3.8). Then, for all solutions of the system (3.9.9), the relationships

$$x(t) = A^{\mathscr{D}}(t)A(t)Z(t)Z^{-1}(s)x(s),$$
$$x(t) = X(t)X^{-1}(s)A^{\mathscr{D}}(s)A(s)x(s), \qquad (3.10.11)$$

are valid, where $Z(t)$ and $X(t)$ are the respective matrixants. In this case a stronger statement than Corollary 3.10.1 holds.

Theorem 3.10.1   If the matrix $(A^{\mathscr{D}}A)^*C^*CA^{\mathscr{D}}A$ is constant, then the inequality

$$D \equiv A^*(A^{\mathscr{D}}A)^*C^*CA^{\mathscr{D}}A + (A^{\mathscr{D}}A)^*C^*C(A^{\mathscr{D}}A)A \leqslant 0 \qquad (3.10.12)$$

is a necessary and sufficient condition for the validity of the inequality

$$\|C(t)x(t)\| \leqslant \|C(s)x(s)\| \qquad (3.10.13)$$

(for all solutions $x(t)$ of the system (3.10.9) and for $t \geqslant s$).

Sufficiency   By multiplying the relationship (3.10.11) by the matrix $E - A^{\mathscr{D}}A$, we obtain $(E - A^{\mathscr{D}}A)x = 0$. But then, putting $M = (A^{\mathscr{D}}A)^*C^*CA^{\mathscr{D}}A$ and $B = E$ into (3.10.6) the relationship (3.10.6) can be written as

$$(Dx', x') = (\|C(t)x(t)\|^2)'. \qquad (3.10.14)$$

From the obtained equality it follows that if $D \leqslant 0$, then $(\|C(t)x(t)\|^2)' \leqslant 0$ and, therefore, when $t \geqslant s$ the inequality (3.10.13) is valid.

Necessity   According to the first formula of (3.10.11), the general solution of the system (3.10.9) is represented on the segment $[\alpha, \beta]$ as

$$x(t) = A^{\mathscr{D}}(t)A(t)Z(t)\gamma,$$

where γ is an arbitrary vector. But then for an arbitrary solution of the system (3.10.9) we obtain:

$$(Dx', x') = ((A^{\mathscr{D}}A)^*C^*CA^{\mathscr{D}}Ax', Ax') + ((A^{\mathscr{D}})^*C^*C(A^{\mathscr{D}}A)Ax', Ax')$$
$$= ((A^{\mathscr{D}}A)^*C^*CA^{\mathscr{D}}x, x) + ((A^{\mathscr{D}})^*C^*CA^{\mathscr{D}}Ax, x) = (C^*CA^{\mathscr{D}}x, x)$$
$$+ (C^*Cx, A_x^{\mathscr{D}}) = (C^*CA^{\mathscr{D}}Z\gamma, A^{\mathscr{D}}AZ\gamma) = C^*CA^{\mathscr{D}}AZ\gamma, A^{\mathscr{D}}Z\gamma). \qquad (3.10.15)$$

In particular, when $\gamma - Z^{-1}A\delta$, where δ is an arbitrary vector, the equality (3.10.15) has the form

$$(Dx', x') = (A^*(A^{\mathscr{D}}A)^*C^*CA^{\mathscr{D}}A\delta, \delta) + ((A^{\mathscr{D}}A)^*C^*C(A^{\mathscr{D}}A)A\delta, \delta) = (D\delta, \delta).$$

Hence it is evident that if the inequality (3.10.13) is satisfied (and, consequently, by virtue of (3.10.14) $(Dx', x') \leqslant 0$), then for an arbitrary vector $\delta (D\delta, \delta \leqslant 0$ and, therefore, $D \leqslant 0$. Thus, the necessity of the condition (3.10.12) is proved.

Corollary 3.10.2 If $CA^{\mathscr{D}}A = C$ and the matrix C^*C is constant, then for all solutions of the system (3.10.9) to satisfy the estimation (3.10.13) (when $t \geqslant s$), it is necessary and sufficient that the inequality $A^*C^*C + C^*CA \leqslant 0$ is satisfied.

Proof If $CA^{\mathscr{D}}A = C$, then (see (3.10.12)) $D = A^*C^*C + C^*CA$ and it remains to refer the reader to the proved Theorem 3.10.1.

Useful generalizations of these results are obtained with the help of the substitution $x(t) = e^{\delta t}y(t)$, where $\delta =$ constant. The following statement, for example, is valid.

Corollary 3.10.3 Let δ be a real number that is not an inverse value of the eigenvalue of the matrix $A(t)$ (for any $t \in [\alpha, \beta]$). Then, if the matrix C^*C is constant and

$$A^*C^*C + C^*CA - 2\delta A^*C^*CA \leqslant 0, \qquad (3.10.16)$$

then for all solutions of equation (3.10.9) when $t \geqslant s$, the estimation

$$\| C(t)x(t) \| \leqslant e^{\delta(t-s)} \| C(s)x(s) \|$$

holds.

Proof In equation (3.10.9) we make the replacement $x(t) = e^{\delta t}y(t)$. Then, for the vector $y(t)$, we obtain:

$$A(E - \delta A)^{-1}y' = y. \qquad (3.10.17)$$

By applying Corollary 3.10.1 to equation (3.10.17) we obtain that if the inequality

$$(E - \delta A^*)^{-1}A^*C^*C + C^*CA(E - \delta A)^{-1} \leqslant 0 \qquad (3.10.18)$$

is satisfied, then for $t \geqslant s$ for all solutions of equation (3.10.17) the estimation

$$\| C(t)y(t) \| \leqslant \| C(s)y(s) \|$$

holds, and for the initial equation (3.10.9) the estimation

$$\| C(t)x(t) \| \leqslant e^{\delta(t-s)} \| C(s)x(s) \|$$

holds.

It remains to make sure that the inequalities (3.10.16) and (3.10.18) are equivalent.

Thus, let inequality (3.10.18) be satisfied, i.e.

$$((E - \delta A^*)^{-1}A^*C^*Cx, x) + (C^*CA(E - \delta A)^{-1}x, x) \leqslant 0$$

for any vector x. By replacing $z = (E - \delta A)^{-1}x$ we obtain:

$$((A^*C^*C + C^*CA - 2\delta A^*C^*CA)z, z) \leqslant 0, \qquad (3.10.19)$$

where the vector z is obviously arbitrary, together with the vector x. But in such a case the inequality (16) is valid. Similarly, by making the inverse substitution in (3.10.19), from inequality (3.10.16) we arrive at the inequality (3.10.18). Corollary 3.10.3 is proved.

It is easy to apply these results when estimating linear combinations of the components of solutions of the inhomogeneous systems (3.10.8). For example, let ind $A(t) \leq 1$. Then, obviously, the matrix $A(t)$ has the property Ω; therefore, on the basis of Theorem 3.7.1 for the system (3.10.8), the general solution

$$x(t) = A^{\mathscr{D}}(t)A(t)Z(t)x(\alpha)$$

$$+ A^{\mathscr{D}}(t)A(t)\int_{\alpha}^{t} Z(t)Z^{-1}(\tau)A^{\mathscr{D}}(\tau)\{E - A'(\tau)[E$$

$$- A^{\mathscr{D}}(\tau)A(\tau)]\}f(\tau)d\tau - [E - A^{\mathscr{D}}(t)A(t)]f(t). \qquad (3.10.20)$$

If in this case the matrix $C(t)$ satisfies the conditions of Corollary 3.10.1, then, by multiplying (3.10.20) by $C(t)$, we obtain:

$$\|C(t)x(t)\| \leq \|C(\alpha)x(\alpha)\|$$

$$+ \int_{\alpha}^{t} \|C(\tau)A^{\mathscr{D}}(\tau)\{E - A'(\tau)[E - A^{\mathscr{D}}(\tau)A(\tau)]\}f(\tau)\| d\tau$$

$$+ \|C(t)[E - A^{\mathscr{D}}(t)A(t)]f(t)\|. \qquad (3.10.21)$$

If the smoothness requirements of the function $f(t)$ are increased, then with the help of the estimation (3.10.21) one can also obtain other estimates which, in the analysis of singular systems, are of even greater significance when compared with estimation (3.10.21). We represent, for example, the solution of equation (3.10.8) in terms of the difference $x = y - f$, the vector y in which satisfies the equation

$$Ay' = y + Af'.$$

Estimation (3.10.21) for vector y has the form

$$\|C(t)y(t)\| \leq \|C(\alpha)y(\alpha)\| + \int_{\alpha}^{t} \|C(\tau)A^{\mathscr{D}}(\tau)A(\tau)f'(\tau)\| d\tau.$$

But, then, in view of the inequalities

$$\|Cx\| \leq \|Cy\| + \|Cf\|, \qquad \|Cy\| \leq \|Cx\| + \|Cf\|,$$

it is easy to obtain an estimate for the vector x, namely

$$\|C(t)x(t)\| \leq \|C(\alpha)x(\alpha)\|$$

$$+ \int_{\alpha}^{t} \|C(\tau)A^{\mathscr{D}}(\tau)A(\tau)f'(\tau)\| d\tau + \|C(\alpha)f(\alpha)\| + \|C(t)f(t)\|. \qquad (3.10.22)$$

As an example, we shall illustrate the perturbation method implying the transition from the singular equation

$$A(t)x'(t) = x(t) + f(t), \quad \alpha \leqslant t \leqslant \beta, \tag{3.10.23}$$

to the equation

$$(A - \varepsilon E)x'_\varepsilon = x_\varepsilon + f, \quad \varepsilon > 0, \tag{3.10.24}$$

with the nonsingular matrix at the derivatives. As for the matrix A, we assume that $A^* + A \leqslant 0$. Under such an assumption, as is known, ind $A(t) \leqslant 1$ and, consequently, the estimation (3.10.22) is applicable. Moreover, by virtue of Corollary 3.10.1, estimation (3.10.22) turns out valid when $C(t) = E$. However, one can also do without this remark. Indeed, let us introduce into our treatment the deviation $v_\varepsilon = x_\varepsilon - x$ and assume that $x_\varepsilon(\alpha) = x(\alpha) = a$ such that $v_\varepsilon(\alpha) = 0$. The deviation v_ε obviously satisfies the equation

$$(A - \varepsilon E)v'_\varepsilon = v_\varepsilon + \varepsilon x', \quad v_\varepsilon(\alpha) = 0, \tag{3.10.25}$$

in which, as a consequence of the inequality $A^* + A - 2\varepsilon E < 0$ and Theorem 2.7.3, the matrix at the derivatives is nonsingular. Moreover, by virtue of Corollary 3.10.1, equation (3.10.25) is satisfied by estimation (3.10.22) (when $C = E$), which in the case of equation (3.10.25) has the form

$$\| v_\varepsilon(t) \| \leqslant \varepsilon \int_\alpha^t \| x''(\tau) \| d\tau + \varepsilon \| x'(\alpha) \| + \varepsilon \| x'(t) \|.$$

From this estimation it follows that if the solution of the initial equation (3.10.23) is sufficiently smooth, then $v_\varepsilon(t) \to 0$ when $\varepsilon \to 0$ (at any fixed $t \in [\alpha, \beta]$).

3.11. Summary of formulae for the solution of the Cauchy problem

If the Cauchy problem is involved, then the formulae for the solution of singular systems constructed in the preceding sections simplify considerably. Our purpose here is to write these formulae for the case of the two following problems:

$$A(t)x'(t) = B(t)x(t) + f(t), \quad \alpha \leqslant t \leqslant \beta,$$
$$x(\alpha) = a, \tag{3.11.1}$$

$$A(t)x'(t) = x(t) + f(t), \quad \alpha \leqslant t \leqslant \beta,$$
$$x(\alpha) = a, \tag{3.11.2}$$

under the assumption that solutions exist and are unique.

First of all, we shall consider problem (3.11.2) (problem (3.11.1) is sometimes reduced to problem (3.11.2) with the help of the substitution $x(t) = e^{ct}y(t)$).

Formula 1:

$$x(t) = A^{\mathscr{D}}(t)A(t)Z(t)A^{\mathscr{D}}(\alpha)A(\alpha)a$$

$$+ A^{\mathscr{D}}(t)A(t) \int_{\alpha}^{t} Z(t)Z^{-1}(\tau)A^{\mathscr{D}}(\tau)[A'(\tau)\zeta(\tau) + f(\tau)]d\tau + \zeta(t),$$

where $Z(t)$ is the solution of the Cauchy problem

$$Z'(t) = \{A^{\mathscr{D}}(t) + [E - A^{\mathscr{D}}(t)A(t)]A'(t)A^{\mathscr{D}}(t)\} Z(t),$$

$$Z(\alpha) = E, \tag{3.11.3}$$

and the vector $\zeta(t)$ is calculated by the formula

$$\zeta(t) = -[E - A^{\mathscr{D}}(t)A(t)]f(t) - (E - A^{\mathscr{D}}A)\sum_{v=1}^{k-1} A^{v}f^{(v)}(t),$$

where k is an integer number satisfying the inequality ind $A(t) \leqslant k$.

This formula is applicable if the matrix $(E - A^{\mathscr{D}}A)A$ is constant (the property Ω). Remember that under such a condition the matrices $(E - A^{\mathscr{D}}A)A^{v}$ $(v = 1, 2, \ldots)$ also are constant.

Formula 2:

$$x(t) = A^{\mathscr{D}}AZ(t)a + A^{\mathscr{D}}A \int_{\alpha}^{t} Z(t)Z^{-1}(\tau)A^{\mathscr{D}}(\tau)f(\tau)d\tau + \zeta(t),$$

where $Z(t)$ is the solution of the Cauchy problem

$$Z'(t) = A^{\mathscr{D}}(t)Z(t), \qquad Z(\alpha) = E, \tag{3.11.4}$$

and the vector $\zeta(t)$ is calculated by the formula

$$\zeta(t) = -(E - A^{\mathscr{D}}A)\sum_{v=0}^{k-1} A^{v}f^{(v)}(t),$$

where $k \geqslant$ ind $A(t)$.

This formula is applicable if the matrices $(E - A^{\mathscr{D}}A)A$ and $A^{\mathscr{D}}A$ are constant (the full property Ω). In this case, as a consequence, the matrices $(E - A^{\mathscr{D}}A)A^{v}$ $(v = 0, 1, 2, \ldots)$ also are constant.

Formula 3:

$$x(t) - A^{\mathscr{D}}(t)A(t)Z(t)A^{\mathscr{D}}(\alpha)A(\alpha)a$$

$$+ A^{\mathscr{D}}(t)A(t) \int_{\alpha}^{t} Z(t)Z^{-1}(\tau)A^{\mathscr{D}}(\tau)\{[A^{\mathscr{D}}(\tau)A'(\tau)A(t)$$

$$+ A'(\tau)]\zeta(\tau) + f(\tau)\} d\tau + \zeta(t),$$

where $Z(t)$ is the solution of the Cauchy problem (3.11.3), and the vector $\zeta(t)$ is calculated by the formula

$$\zeta(t) = -[E - A^{\mathscr{D}}(t)A(t)]f(t) - D(t)\{E + [E$$
$$- A^{\mathscr{D}}(t)A(t)]A'(t)\}^*[E - A^{\mathscr{D}}(t)A(t)]A(t)f'(t),$$

where

$$Z'(t) = \{A^{\mathscr{D}}(t) + [E - A^{\mathscr{D}}(t)A(t)]A'(t)A^{\mathscr{D}}(t)\}Z(t). \tag{3.11.5}$$

This formula is applicable if ind $A(t) \leqslant 2, (E - A^{\mathscr{D}}A)AA'A^{\mathscr{D}} = 0,$ and the inverse matrix (3.11.5) exists on the entire segment $[\alpha, \beta]$.

Formula 4:

$$x(t) = A^{\mathscr{D}}AZ(t)a + A^{\mathscr{D}}A\int_\alpha^t Z(t)Z^{-1}(\tau)A^{\mathscr{D}}(\tau)f(\tau)\alpha\tau + \zeta(t),$$

where $Z(t)$ is the solution of the problem (3.11.4), and the vector $\zeta(t)$ is calculated by the formula

$$\zeta(t) = -(E - A^{\mathscr{D}}A)f(t) - D(t)[E - A'(t)]^*(E - A^{\mathscr{D}}A)A(t)f'(t),$$

in which

$$D = [A^*A + (E + A')^*(E + A')]^{-1}. \tag{3.11.6}$$

This formula is applicable if ind $A(t) \leqslant 2,$ the matrix $A^{\mathscr{D}}A$ is constant, and, moreover, the inverse matrix (3.11.6) exists on the entire segment $[\alpha, \beta]$.

We now turn to the problem (3.11.1).

Formula 5:

$$x(t) = Z(t)a + \int_\alpha^t Z(t)Z^{-1}(\tau)A^+(\tau)f(\tau)\,d\tau,$$

where $Z(t)$ is the solution of the Cauchy problem

$$Z'(t) = A^+(t)B(t)Z(t), \qquad Z(\alpha) = E.$$

This formula is applicable if the columns of the matrix $A(t)$ are linearly independent (in such a case $A^+ = (A^*A)^{-1}A^*$).

The problem (3.11.1) can also be solved with the help of formulae 1–4 if the columns of the matrix $B(t)$ are linearly independent and there exists a resolving pair of matrices (A^B, Y) satisfying the requirements formulated in Section 3.6.

The obtained formulae are easily applied when constructing difference schemes. Let us demonstrate this in the example of the system (3.11.2) having the full property Ω.

In this case the solution $x(t)$ of problem (3.11.2) can be obtained by Formula (3.11.2) and, consequently, as is easy to see,

$$x(t) = A^{\mathscr{D}}Ay(t) + \zeta(t), \tag{3.11.7}$$

where the vector $y(t)$ is the solution of the problem

$$y'(t) = A^{\mathscr{D}}(t)y(t) + A^{\mathscr{D}}(t)f(t),$$

$$y(\alpha) = a, \quad \alpha \leqslant t \leqslant \beta, \tag{3.11.8}$$

the vector $\zeta(t)$ is calculated by the formula

$$\zeta(t) = -(E - A^{\mathscr{D}}A) \sum_{v=0}^{k-1} A^v f^{(v)}(t), \tag{3.11.9}$$

and the matrices $A^{\mathscr{D}}A$ and $(E - A^{\mathscr{D}}A)A^v$ $(= 0, 1, \dots, k-1)$ are constant. By applying, for example, the implicit Euler scheme to the solutions of the problem (3.11.8) and replacing, after that, the matrices $A^{\mathscr{D}}, A^{\mathscr{D}}A$, and $(E - A^{\mathscr{D}}A)$ in (3.11.7)–(3.11.9) with their approximations, we obtain:

$$(A_i^{k+1} - \tau A_i^k + \tau^2 E)y_i = (A_i^{k+1} + \tau^2 E)y_{i-1} + \tau A_i^k f(i\tau), \quad i = 1, \dots, N,$$

$$N\tau = \beta - \alpha, \qquad y_0 = a, \tag{3.11.10}$$

$$(A_i^{k+1} + \tau^2 E)x_i = A_i^{k+1} y_i - \tau^2 \sum_{v=0}^{k-1} A_i^v f^{(v)}(i\tau),$$

$$A_i = A(i\tau)$$

and the equality (3.11.10), in view of the full property Ω, can also be written as

$$(A_0^{k+1} + \tau^2 E)x_i = A_0^{k+1} y_i - \tau^2 \sum_{v=0}^{k-1} A_0^v f^{(v)}(i\tau).$$

We leave to the reader the proof of the convergence of x_i and $x(i\tau)$.

Other information about the properties of generalized inverse matrices and their applications to the solution of singular systems can be taken from [28–37].

3.12. Extended systems with differential connection

Let us attach to the equation

$$\Phi(x'(t), x(t), t) = 0, \quad \alpha \leqslant t \leqslant \beta, \tag{3.12.1}$$

an additional system of differential relationships,

$$F_s(x^{(s)}(t), x^{(s-1)}(t), \dots, x'(t), x(t), t) = 0, \quad s = 1, \dots, k, \tag{3.12.2}$$

and let us differentiate equation (3.12.1) $k - 1$ times in t. As a result, by introducing the designations

$$x^{(i)} = p_i,$$

we obtain the system:

$$p_i' = p_{i+1}, \quad i = 0, \dots, k-1, \tag{3.12.3}$$

$$G_j(p_{j+1}, p_j, \dots, p_0, t) = 0, \quad j = 0, \dots, k-1, \tag{3.12.4}$$

$$F_s(p_s, p_{s-1}, \dots, p_0, t) = 0, \quad s = 1, \dots, k. \tag{3.12.5}$$

Following Yanenko [38], the system (3.12.3), (3.12.4) obtained by differentiat-ing equation (3.12.1) will be called the *extended* system. The final connections (3.12.5) in the space of variables p_i of an extended system define some surface.

The problem implies choosing this surface such that it contains integral varieties of an extended system with a given arbitrariness (see [38]).

The application and development of this idea (in the example of quasi-linear systems with particular derivatives) is constained in a book [39, p. 109], where important results are presented on this issue.

As far as singular systems of ordinary differential equations are concerned, the problem of studying extended systems with differential connections has not yet been formulated properly, although there were some attempts to apply, for example, methods of spline-colocation, which are known to assume introducing connections of the type $x^{(n)}(t) \equiv 0$. Meanwhile, in this case the statement of the problem of investigating an extended system is a very natural one because solutions of singular systems depend, generally speaking, on derivatives of the right-hand side of the system.

The most modest of the problems implies here the following: find differential connections for which the solution of the singular system,

$$A(t)x'(t) = B(t)x(t) + f(t), \quad \alpha \leqslant t \leqslant \beta, \tag{3.12.6}$$

can be obtained using the formulae of the preceding section.

It is easy to obtain a formal solution of this problem. However, there arises the complicated problem of establishing the compatibility of the differential connections introduced, with the system (3.12.6). Let us consider an example.

Let it be necessary to find a connection, for which the system (3.12.6) allows a solution of the form (3.3.17) in the case of the arbitrary matrices A and B.

To solve this problem, we shall use Lemma 3.3.1, according to which equation (3.12.6) is equivalent to the system

$$x' = A^- Bx + A^- f + (E - A^- A)x', \tag{3.12.7}$$

$$(E - AA^-)Bx = -(E - AA^-)f. \tag{3.12.8}$$

It is obvious that if the vector $(E - A^- A)x'$ is specified, then equation (3.12.7) becomes explicit, and the system (3.12.7), (3.12.8) is easily solved with the help of Theorem 3.3.2. Thus, as the differential connection, it is necessary to take the connection $(E - A^- A)x' = (E - A^- A)\varphi$, where φ is a given vector.

As a result, we arrive at the problem of determining a general solution of the system

$$Ax' = Bx + f, \tag{3.12.9}$$

$$(E - A^- A)x' = (E - A^- A)\varphi, \quad \alpha \leqslant t \leqslant \beta, \tag{3.12.10}$$

which, as is easily verified, is equivalent to the system

$$x' = A^- Bx + A^- f + (E - A^- A)\varphi, \tag{3.12.11}$$

$$(E - AA^-)Bx = -(E - AA^-)f \qquad (3.12.12)$$

(cf. the system (3.3.4)–(3.3.6)).

By repeating now the derivation preceding the formulation of Theorems 3.3.1–3.3.3, in view of the fact that when obtaining a general solution of the system (3.12.9), (3.12.10) for the condition (3.3.2), one should put $C(t) \equiv 0$ and $a = 0$, we arrive at the following results.

Theorem 3.12.1   The problem (3.12.9), (3.12.10) is solvable if and only if at some point $t_0 \in [\alpha, \beta]$ the identity

$$G(s)\Gamma_1^- \Gamma_2(t_0) = G(s)\Phi(s, t_0) - [E - A(s)A^-(s)]f(s), \quad \alpha \leqslant s \leqslant \beta,$$

is satisfied, where

$$G(s) = [E - A(s)A^-(s)]B(s)X(s),$$

the matrix $X(s)$ is the solution of the problem

$$X'(s) = A^-(s)B(s)X(s), \qquad X(\alpha) = E, \qquad (3.12.13)$$

$$\Phi(s, t) = \int_s^t X^{-1}(\tau)\{A^-(\tau)f(\tau) + [E - A^-(\tau)A(\tau)]\varphi(\tau)\}\, d\tau,$$

$$\Gamma_1 = \int_\alpha^\beta G^*(s)G(s)\, ds, \qquad (3.12.14)$$

$$\Gamma_2(t_0) = \int_\alpha^\beta \{G^*(s)G(s)\Phi(s, t_0) - G^*(s)[E - A(s)A^-(s)]f(s)\}\, ds. \quad (3.12.15)$$

Theorem 3.12.2   The general solution of the problem (3.12.9), (3.12.10) (if it exists) can be represented as

$$x(t) = X(t)X^{-1}(t_0)x(t_0) + \int_{t_0}^t X(t)X^{-1}(\tau)\{A^-(\tau)f(\tau) + [E - A^-(\tau)A(\tau)]\varphi(\tau)\}\, d\tau,$$

$$x(t_0) = X(t_0)\Gamma_1^- \Gamma_2(t_0) + X(t_0)\gamma,$$

where t_0 is an arbitrary fixed point of the segment $[\alpha, \beta]$, the matrix $X(t)$ is the solution of the problem (3.12.13), γ is an arbitrary solution of the system

$$\Gamma_1 \gamma = 0, \qquad (3.12.16)$$

and Γ_1 and $\Gamma_2(t_0)$ are calculated by formulae (3.12.14) and (3.12.15).

Theorem 3.12.3   The problem (3.12.9), (3.12.10) has only one solution if and only if $\gamma = 0$ is the unique solution of equation (3.12.16).

Thus, if the differential connection $(E - A^-A)x' = (E - A^-A)\varphi$ is fixed, then the general solution of equation (3.12.9) perhaps will possess only a constant arbitrariness. At the same time, it should be noted that, if the pair of matrices

$(A(t), B(t))$ is perfect, then the vector φ in the differential connection is arbitrary and, therefore, the general solution of equation (3.12.9) can contain both a constant and a functional arbitrariness.

3.13 Nonlinear systems

Considering implicit equations of the general form

$$f(x', x, t) = 0, \quad \alpha \leqslant t \leqslant \beta, \tag{3.13.1}$$

encounters significant difficulties. In this connection, it suffices to note that, even in the linear case, confidence in the validity of the solution algorithms arises if only the pair of matrices in the system has the property Ω or the index of the system does not exceed two. As far as nonlinear equations of the form (3.12.1) are concerned, the difficulties become even more complicated because at present there is not a sufficiently well-developed theory of the existence of an implicit function as a whole (on the entire segment $[\alpha, \beta]$).

At the same time, it was noticed that considerable information about the solutions of equation (3.13.1) is contained in the pair of matrices in the linear equation

$$A(t)x'(t) = B(t)x(t) + A(t)\theta'(t) + B(t)\theta(t), \tag{3.13.2}$$

where $\theta(t)$ is the solution of equation (3.13.1), and the matrices $A(t)$ and $B(t)$ are obtained by the formulae

$$A(t) = f'_{x'}(\theta'(t), \theta(t), t), \qquad B(t) = -f'_x(\theta'(t), \theta(t), t),$$

i.e. are values of the respective Jacobians on the solution of equation (3.13.1).

Despite the nonconstructive character of using equation (3.13.2) to study and solve equation (3.13.1), such an approach sometimes leads not only to an understanding of the properties of a numerical solution but also to establishing theorems of the existence of uniqueness as well.

In this direction (or close to it) a number of results have been obtained during the past five years. We shall note here only some of them [40–49].

In [40], a local theorem of the existence and uniqueness was formulated for solutions of the problem

$$A(t)x'(t) = f(x, t), \qquad \text{rank } A(t) = \text{constant}, \tag{3.13.3}$$

in the case when the bundle $\lambda A(\alpha) - f'_x(a, \alpha)$ satisfies the conditions for the 'rank-degree' criterion, and the compatibility condition

$$\text{rank}[A(\alpha)] = \text{rank}[A(\alpha), f(a, \alpha)]$$

is satisfied.

Also, the convergence of the linearization method (the modified Newton method) is established if in a uniform metric the initial approximation is chosen to be sufficiently close to the solution.

Later, Chistyakov extended these results. He showed that the implicit Euler scheme

$$A(t_{i+1})\frac{y_{i+1}-y_i}{\tau}=f(y_{i+1},t_{i+1}),\qquad y_0=a, \qquad (3.13.4)$$

is converging. It appeared that when the Newton method is used to solve equation (3.13.4) with sufficiently small τ, one can take y_i as the initial approximation for y_{i+1} (as regards the general equation (3.13.1), this fact, generally speaking, does not occur).

A study was also made of the question of discretizing linear systems that result when the problem (3.13.3) is linearized and of the question of the influence of perturbations of input data upon the numerical process.

The 'rank-degree' criterion used in [40] was proposed by Chistyakov and implies the following: the system (3.13.2) satisfies the conditions of this criterion if the degree of the polynomial $\det[B(t)-\lambda A(t)]$ for all $t\in[\alpha,\beta]$ coincides with the rank of the matrix $A(t)$ which, according to the assumption, is constant.

The 'rank-degree' criterion separates the systems (3.13.2), in which the pair of matrices $(A(t),B(t))$, for all $t\in[\alpha,\beta]$, has a simple structure in zero eigenvalues (a canonical representation of the pair of matrices $(A(t),B(t))$ satisfying this condition does not contain nilpotent blocks of the degree above unity).

A large series of interesting papers (see [17, 43–45] and references therein) is due to März, in which for equation (3.13.1) the existence and uniqueness conditions are formulated for the solutions as well as substantiating, with the constraints formulated in the papers just cited, some one- and many-step methods of numerical integration. Multi-step methods for solving equation (3.13.3) were also studied in [46].

It is interesting to note that results obtained by different authors and outwardly not resembling each other show an internal, deep likelihood: the conditions, under which the study is made, are mainly associated with the requirement for simplicity of the structure of the matrices in equation (3.13.2) (see, for example, condition A4 in [43]), which, in our terminology, is equivalent to the validity of the 'rank-degree' criterion conditions or, in other words, that the system (3.13.2) belongs to the class of systems of index 1.

Thus, the most well-studied nonlinear systems are presently those whose index is unity.

3.14. Systems of equations with partial derivatives

Using generalized inverse matrices it is possible to construct a number of useful relationships which must be satisfied by all sufficiently smooth solutions of a system of the form

$$Bu_t+Au_x=f, \qquad (3.14.1)$$

in which A, and B are constant numerical matrices.

We first consider an equation of the particular form

$$u_t + Au_x = f \qquad (3.14.2)$$

by assuming that it is solvable and that $u(x, t)$ is its solution.

Let $\lambda_1, \ldots, \lambda_m$ be all eigen-numbers of the matrix A differing from each other. We generate the matrices $A_i = A - \lambda_i A$ $(i = 1, \ldots, m)$ and represent equation (3.14.2) as

$$u_t + \lambda_i u_x = f - A_i u_x. \qquad (3.14.3)$$

By applying now the method of characteristics to equation (3.14.3) (see, for example, [39]), we obtain the relationship:

$$u(x, t) = u(x - \lambda_i t, 0) + \int_0^t f(x - \lambda_i(t - \tau), \tau) \, d\tau - \int_0^t A_i u_x(x - \lambda_i(t - \tau), \tau) \, d\tau,$$

which will be written as:

$$u(x, t) = T_i u(x, t) + \int_0^t f(x - \lambda_i(t - \tau), \tau) \, d\tau + u(x - \lambda_i t, 0), \qquad (3.14.4)$$

where T_i is an operator the action of which on the function $\chi(x, t)$ is defined by the formula

$$T_i \chi(x, t) = - \int_0^t A_i \chi_x(x - \lambda_i(t - \tau), \tau) \, d\tau \qquad (3.14.5)$$

(it is convenient to write $T_{i\chi}^0(x, t) = \chi(x, t)$).

By integrating the equality (3.14.4) $r - 1$ times, in view of formula (3.14.5), we obtain:

$$u(x, t) = T_i^r u(x, t) + \sum_{s=0}^{r-1} (-1)^s \frac{f^s}{s!} \frac{\partial^s}{\partial x^s} [A_i^s u(x - \lambda_i t, 0)]$$

$$+ \sum_{s=0}^{r-1} T_i^s \left[\int_0^t f(x - \lambda_i \tau, \tau) \, d\tau \right]. \qquad (3.14.6)$$

We now make use of the equalities

$$\sum_{i=1}^m (E - A_i^{\mathcal{D}} A_i) = E, \qquad (3.14.7)$$

$$(E - A_i^{\mathcal{D}} A_i) A_i^{\nu_i} = 0, \qquad (3.14.8)$$

where ν_i is the index of the matrix A_i (see (1.8.21) and Definition 1.8.1).

By multiplying the equality (3.14.6) to the matrix $E - A_i^{\mathcal{D}} A_i$, and as a consequence of property (3.14.8) and formula (3.14.5) when $r = \nu_i$, we obtain:

$$(E - A_i^{\mathcal{D}} A_i) u(x, t) = \sum_{s=0}^{\nu_i - 1} (-1)^s \frac{t^s}{s!} (E - A_i^{\mathcal{D}} A_i) A_i^s \frac{\partial^s}{\partial x^s} u(x - \lambda_i t, 0)$$

$$+ \sum_{s=0}^{\nu_i - 1} (E - A_i^{\mathcal{D}} A_i) T_i^s \left[\int_0^t f(x - \lambda_i \tau, \tau) \, d\tau \right]. \qquad (3.14.9)$$

By summing the equalities (3.14.9) in $i = 1, \ldots, m$ and also taking property (3.14.7) into account, we arrive at the main relationship:

$$u(x, t) = \sum_{i=1}^{m} (E - A_i^{\mathcal{D}} A_i) \sum_{s=0}^{v_i - 1} (-1)^s \frac{t^s}{s!} A_i^s \frac{\partial^s}{\partial x^s} u(x - \lambda_i + 0)$$

$$+ \sum_{i=1}^{m} (E - A_i^{\mathcal{D}} A_i) \sum_{s=0}^{v_i - 1} T_i^s \left[\int_0^t f(x - \lambda_i \tau, \tau) \, d\tau \right]. \qquad (3.14.10)$$

Let us now turn to considering equations of the general form (3.14.1) by assuming that the pair of matrices (B, A) is imperfect in it. This assumption is not substantial because after the replacement $u(x, t) = v(x, \tau)$, where $\tau = t - cx$, $c = $ constant, the pair of matrices in the system becomes perfect (for almost all c, except, perhaps, only for their final number; see Theorem 1.6.2 and Remark 1.6.1).

If the vector u is the solution of the system (3.14.1), then according to the properties of semi-inverse matrices there exists such a vector v that the pair (u, v) satisfies the system

$$u_t + B^- A u_x = B^- f + v, \qquad (3.14.11)$$

$$Bv = 0, \qquad (3.14.12)$$

$$(E - BB^-) A u_x = (E - BB^-) f \qquad (3.14.13)$$

(see Corollary 1.3.1).

Conversely, if the pair (u, v) is the solution of the system (3.14.11)–(3.14.13), then, as can easily be verified, the vector u from it is the solution of equation (3.14.1).

Thus, the solution of equation (3.14.1) is reduced to solving the system (3.14.11)–(3.14.13), which allows us—when obtaining the relationship for equation (3.14.1)—to use formula (3.14.10).

By applying this formula to equation (3.14.11) we obtain the required relationship:

$$u(x, t) = \sum_{i=1}^{m} (E - A_i^{\mathcal{D}} A_i) \sum_{s=0}^{v_i - 1} (-1)^s \frac{t^s}{s!} A_i^s \frac{\partial^s}{\partial x^s} u(x - \lambda_i t, 0) + \sum_{i=1}^{m} (E - A_i^{\mathcal{D}} A_i)$$

$$\times \sum_{s=0}^{v_i - 1} T_i^s \left\{ \int_0^t [B^- f(x - \lambda_i \tau, \tau) + v(x - \lambda_i \tau, \tau)] \, d\tau \right\}, \qquad (3.14.14)$$

where λ_i $(i = 1, \ldots, m)$ are all (differing from each other) eigen-numbers of the matrix $B^- A$, $A_i = B^- A - \lambda_i E$, and v_i is the index of matrix A_i. The operator T_i is determined by formula (3.14.5).

We now substitute (3.14.14) into (3.14.13). As a result, taking into account the perfectness of the pair of matrices (B, A) (see (1.6.1)) as well as Theorem 1.8.4

and the equality $v = (E - B^- B)v$, we obtain:

$$(E - BB^-)A \sum_{i=1}^{m} (E - A_i^{\mathscr{D}} A_i) \sum_{s=0}^{v_i - 1} \left\{ (-1)^s \frac{t^s}{s!} A_i^s \frac{\partial^{s+1}}{\partial x^{s+1}} u(x - \lambda_i t, 0) \right.$$

$$\left. + \frac{\partial}{\partial x} T_i^s \left[\int_0^t B^- f(x - \lambda_i \tau, \tau) \, d\tau \right] \right\} = (E - BB^-)f(x, t). \tag{3.14.15}$$

If the Cauchy problem with an initial given $u(x, 0) = \varphi(x)$ is dealt with, then equality (3.14.15) is the matching condition for the initial data with the right-hand side. For example, when $t = 0$, we have

$$(E - BB^-)A\varphi'(x) = (E - BB^-)f(x, 0).$$

A full set of matching conditions is obtainable by differentiating the equality (3.14.15) in t and with the substitution of $t = 0$.

Other variations on this subject are contained in [50] (see also [51]).

REFERENCES

[1] Gantmakher F. R. (1966) *Theory of Matrices* Moscow: Nauka.
[2] Boyarintsev Yu. E. (1980) *Regular and Singular Systems of Linear Ordinary Differential Equations* Novosibirsk: Nauka.
[3] Albert A. (1977) *Regression, Pseudo-inversion and Recurrent Evaluation* Moscow: Nauka.
[4] Glazman I. M. and Lyubich Yu. I. (1969) *Finite-dimensional Linear Analysis in Problems* Moscow: Nauka.
[5] Korsukov V. M. (1978) *One Theorem on Rank of the Matrix Product and Some Related Corollaries* Prikladnaya matematika—Irkutsk: SEI CO AN SSSR 154–159.
[6] Rao C. R. and Mitra S. K. (1971) *Generalized Inverse of Matrices and its Applications* New York: Wiley.
[7] Godunov S. K. and Rabenky V. S. (1963) Canonical forms of systems of linear ordinary difference equations with constant coefficients *Zhurn. vychisl. matematiki i mat. fiziki*, **3**(2), 211–222.
[8] Wilkinson J. H. (1977) *The Differential System Bx′ = Ax and Generalized Eigenvalue Problem Au = Bu* Nat. Phys. Lab., Rep. NAC 73.
[9] Boyarintsev Yu. E. (1981) *On the Theory of Systems with Rectangular Matrices of Coefficients* Chislennye metody optimizatsii i ikh prilozheniya, Irkutsk: SEI SO AN SSSR, 106–117.
[10] Boyarintsev Yu. E. (1982) *On systems of ordinary differential equations unsolved for derivatives* Vyrozhdennye sistemy obyknovennykh differentsialnykh uravneniy, Novosibirsk: Nauka, 5–19.
[11] Beckenbach E. and Bellman R. (1965) *Inequalities* Moscow: Mir.
[12] Drasin M. P. (1958) Pseudoinverses in associative rings and semigroups *Amer. Math. Monthly* **65**, 506–515.
[13] Carl D. and Meyer Jr (1974) Limits and the index of a square matrix *SIAM J. Appl. Math.* **26** (3) 469–478.
[14] Boyarintsev Yu. E. (1981) *The Resolving Pair of Matrices* Priblizhennye metody resheniya operatornykh uravneniy i ikh prilozheniya, Irkutsk: SEI SO AN SSSR, 35–47.
[15] Boyarintsev Yu. E. (1983) *A Resolving Pair and Its Applications* Aktualnye problemy vychislitelnoi i primladnoi matematiki, Novosibirsk: Nauka.
[16] Boyarintsev Yu. E. (1983) *On Representation of Solutions of a System of Linear Algebraic Equations With the Help of Generalized Inverse Matrices* Vychislitelnye metody lineinoi algebry, Moscow: Otdel vychislitelnoi matematiki AN SSSR, 33–45.

[17] März R. (1984) On difference and shooting for boundary value problems in differential algebraic equations *Z. Angew. Math. Mech.* **64** (11), 463–473.

[18] Chistyakov V. F. (1982) *Application of Difference Methods for Solving Linear Systems not Resolved for the Derivative* Metody optimizatsii i ikh prilozheniya, Irkutsk: SEI SO AN SSSR, 145–149.

[19] Chistyakov V. F. (1984) On perturbation of quasilinear systems of ordinary differential equations with a degenerate matrix at derivatives *Chislennye metody mekhaniki sploshnoi sredy*, *Novosibirsk* **15** (5), 154–166.

[20] Boyarintsev Yu. E. (1986) *Degenerate Systems and the Index of a Variable Matrix* Differentsialnye uravneniya i chislennye metody, Novosibirsk: Nauka, 105–114.

[21] Samarsky A. A. (1977) *The Theory of Difference Schemes* Moscow: Nauka.

[22] Samarsky A. A. and Gulin A. V. (1973) *Stability of Difference Schemes* Moscow: Nauka.

[23] Filippov A. F. (1985) *Differential Equations with the Discontinuous Right-hand Side* Moscow: Nauka.

[24] Chistyakov V. F. (1979) *On the Solution of Linear Singular Systems of Ordinary Differential Equations with a Constant Coefficient by the Method of Eliminating unknowns* Metody optimizatsii i ikh prilozheniya, Irkutsk: SEI SO AN SSSR, 160–165.

[25] Boyarintsev Yu. E., Loginov A. A., Chistyakov V. F., and Fedchenko Z. A. (1983) *An Applied Program Package for Numerical Integration of Singular Systems of Ordinary Differential Equations* Algoritmy i programmy/Gos. fond algoritmov i programm, Moscow: VNTITSentr, No. 6(57), 40.

[26] Danilov V. A. (1983) *Causes of Difficulties of Numerical Integration of Some Rigid Systems of Ordinary Differential Equations Close to Degenerate Ones* Dinamika nelineinykh sistem, Novosibirsk: Nauka, 173–182.

[27] Marchuk G. I. and Lebedev V. I. (1971) *Numerical Methods in the Neutron Transfer Theory* Moscow: Atomizdat.

[28] Boyarintsev Yu. E. and Korsukov V. M. (1975) *Application of Difference Methods to Solving Regular Systems of Ordinary Differential Regular Equations* Voprosy prikladnoi matematiki Irkutsk: SEI SO AN SSSR, 140–158.

[29] Boyarintsev Yu. E. and Korsukov V. M. (1977) *Structure of a General Continuously Differentiable Solution of Boundary Value Problem for a Singular System of Ordinary Differential Equations.* Voprosy prikladnoi matematiki, Irkutsk: SEI SO AN SSSR, 73–93.

[30] Boyarintsev Yu. E. (1976) *On the Theory of Boundary-value Problem for Systems of Ordinary Differential Equations* Metody optimizatsii i issledovaniye operatsiy (prikladnaya matematika), Irkutsk: SEI SO AN SSSR, 89–104.

[31] Boyarintsev Yu. E. (1977) *On general solutions of boundary-value problems for singular systems of ordinary differential equations* Chislennye metody mekhaniki sploshnoi sredy, *Novosibirsk* **8** (7), 12–21.

[32] Boyarintsev Yu. E. (1978) *On the Structure of a General Solution of the Boundary-value Problem for Singular Systems of Ordinary Differential Equations* Novosibirsk: Computing Center SO AN SSSR, Preprint 44.

[33] Boyarintsev Yu. E. (1978) *Modular Analysis of a Computational Algorithm for Solving Boundary-value Problems for Linear Singular Systems of Ordinary Differential Equations* Kompleksy programm matematicheskoi fiziki (materialy V Vsesoyuznogo seminara po kompleksam programm matematicheskoi fiziki), Novosibirsk: ITPM SO AN SSSR, p. 3–14.

[34] Boyarintsev Yu. E. (1978) *Singular Systems of Linear Ordinary Differential Equations and Numerical Methods of Solving Them* Prikladnaya matematika, Novosibirsk: Nauka, 72–108.

[35] Boyarintsev Yu. E. (1978) *On one Representation of the Inverse Drasin Matrix* Chislennye metody optimizatsii (prikladnaya matematika), Irkutsk: SEI SO AN SSSR, 176–179.

[36] Boyarintsev Yu. E. and Boyarintseva T. P. (1983) *A Remark on an Implicit Difference Scheme Approximating the System of Stokes Equations* Chislennye metody analiza i ikh prilozheniya, Irkutsk: SEI SO AN SSSR, 127–131.

[37] Boyarintsev Yu. E. (1986) *Application of the Inverse Drasin Matrix for Constructing Stable Difference Approximations to Degenerate Systems of Linear Ordinary Differential Equations* Irkutsk: Irkutsk Computing Center SO AN SSSR, Preprint No. 4.

[38] Yanenko N. N. (1964) *The Theory of Comptability and Methods of Integrating Systems of Nonlinear Equations in Partial Derivatives* Trudy IV Vsesoyuznogo matematicheskogo syezda, Leningrad: Nauka, 247–252.

[39] Rozhdestvensky B. L. and Yanenko N. N. (1978) *Systems of Quasilinear Equations and Their Applications to Gas Dynamics* Moscow: Nauka.

[40] Chistyakov V. F. (1982) *On Linearization of Degenerate Systems of Quasilinear Ordinary Differential Equations and Their Applications* Irkutsk: SEI SO AN SSSR 146–157.

[41] Chistyakov V. F. (1983) *On Properties of Quasilinear Degenerate Systems of Ordinary Differential Equations* Dinamika nelineinykh sistem, Novosibirsk: Nauka, 164–173.

[42] Chistyakov V. F. (1984) *On the Solution Existence for Degenerate Quasilinear Systems of Ordinary Differential Equations* Chislennye metody analiza i ikh prilozheniya, Irkutsk: SEI SO AN SSSR, 151–159.

[43] März R. (1984) *On numerical Methods for BVP's in Transferable DAE's* Berlin: Sektion Mathematik der Humboldt-Universität zu Berlin, Seminarbericht No. 65, 145–148.

[44] März R. (1984) *On Correctness and Numerical Treatment of Boundary Value Problems in DAE's* Berlin: Sektion Mathematik, Preprint No. 73.

[45] März R. (1985) On initial value problems in differential-algebraic equations and their numerical treatment. *Computing* **35**, 13–37.

[46] Griepentrog E. (1984) *Numerical Integration of Transferable Implicit Systems* Berlin: Sektion Mathematik der Humboldt-Universität zu Berlin, Seminararbeit No. 65, 49–59.

[47] Gear C. W. and Petzold L. R. (1966) *ODE Methods for the Solutions of Differential/Algebraic System*, SAND82-8051, Sandia National Laboratories, Livermore, and *SIAM J. Numer. Anal.* **21** (4), 716–728.

[48] Petzold L. R. (1984) *Solution of Nonlinear Differential/Algebraic Systems by Numerical ODE Methods* Berlin: Sektion Mathematik der Humboldt-Universität zu Berlin, Seminarbericht, No. 65, 161–172.

[49] Campbell S. L. and Petzold L. R. (1983) Canonical forms and solvable singular systems of differential equations *SIAM J. Alg. and Discrete Methods* **4**(4), 517–521.

[50] Boyarintsev Yu. E. (1984) *Application of Generalized Inverse Matrices to Solving and Investigating Differential Equations with First-order Partial Derivatives* Metody optimizatsii i issledovaniye operatsii, Irkutsk: SEI SO AN SSSR 123–141.

[51] Boyarintsev Yu. E. (1983) Application of the inverse Drasin matrix to the solution of degenerate systems with first-order partial derivatives *Chislennye metody mekhaniki sploshnoi sredy, Novosibirsk* **14** (6), 27–30.

INDEX

Index compiled by Paul Nash